Published by
Princeton Architectural Press
202 Warren Street, Hudson, NY 12534
Visit our website at www.papress.com

Published in agreement with
Pavilion Books Company Ltd.
43 Great Ormond Street
London WC1N 3HZ

For Princeton Architectural Press:
Editor: Parker Menzimer
Cover design: Paul Wagner

Special thanks to: Paula Baver, Janet Behning,
Abby Bussel, Jan Cigliano Hartman, Susan Hershberg,
Kristen Hewitt, Stephanie Holstein, Lia Hunt,
Valerie Kamen, Jennifer Lippert, Sara McKay,
Wes Seeley, Sara Stemen, Marisa Tesoro,
and Joseph Weston of Princeton Architectural Press
—Kevin C. Lippert, publisher

Library of Congress Cataloging-in-Publication Data
Names: Barker, Sarah (Television producer), author. |
    Nilsson, Maria, 1979 June 7- illustrator.
Title: 50 things to see in the sky / Sarah Barker ;
    illustrated by Maria Nilsson.
Other titles: Fifty things to see in the sky
Description: New York : Princeton Architectural Press,
    [2019] | Copyrighted by Pavilion Books Company.
Identifiers: LCCN 2018060084 | ISBN
    9781616898007 (hardcover)
Subjects: LCSH: Astronomy—Popular works. |
    Astronomy—Observers' manuals. | Amateur
    astronomy.
Classification: LCC QB44.3 .B27 2019 | DDC
    520—dc23
LC record available at https://lccn.loc.
    gov/2018060084

IMPORTANT SAFETY NOTICE
Never underestimate the power of the Sun.
You can permanently damage your eyes by looking into
the Sun without proper protection. If you are using
a telescope to look at the Sun, be absolutely sure
to use a filter designed for solar viewing. Children
should be supervised while using telescopes, especially
during the daytime.

50 THINGS
TO SEE
IN THE SKY

Sarah Barker

Illustrated by Maria Nilsson

PRINCETON ARCHITECTURAL PRESS · NEW YORK

Contents

Introduction

The sky above us holds limitless wonder. There is an endless array of things to see in the sky, both during the daytime, and particularly at night. People can spend a lifetime looking into the heavens, entire careers studying the science of the skies, and *still* be surprised by what's out there.

Collectively, we have been amazed by astronomy for as long as we've been human. And perhaps even longer. Throughout history (and even pre-history) civilizations from every corner of the world have gazed into the starlit skies. Names, stories, and significance have been granted to groups of stars—the constellations—since time immemorial. Astronomy may be the oldest profession on Earth.

Before we had smartphones and tablets, computers and televisions, we would look to the heavens for entertainment. Written in the constellations were magical creatures, epic battles, great loves won and lost. And more than just providing amusement, the shapes of the stars were used for navigation, their predictable procession through the sky serving as markers for when to plant or harvest crops, while the phases of the Moon measured the passing of time.

In our modern society, with its many distractions down on the ground, we may be more prone to forget the shining stars above. Certainly our bright lights and sparkling cities have made the nights much lighter than they were to our ancestors, and some things are more difficult to see. But stars still shine overhead, comets and meteors come into view, and the planets wander through the sky. In fact, there is much that is visible to us now that once we could have only dreamed of seeing: the birthplaces of stars, entire groups of galaxies, even planets in other solar systems.

And it's not as hard to see some of these spectacles as you might think! You don't need all the latest gear, you don't need to spend all your money on telescopes and high-tech kit, often you just need your eyes and a clear, dark sky.

This book is divided into three sections: things that don't require any equipment to see, things that are a little more challenging to spot without a telescope, and a few things that are trickier still to view. Some objects in this book can only be seen from the northern hemisphere, some only from the southern hemisphere, and some only with the help of professional telescopes (or looking up images online!). Most, however, are available to us all, at some point or another, and even if you can't get to see them for yourself, I hope the explanations and illustrations shown here are sufficient to trigger a sense of awe and wonder at what lies in our celestial sphere.

So get out there and look up! Even from the busiest cities in the world, there will be things here that you can see, such as shooting stars, red supergiants, and the International Space Station! But I encourage you to take a road trip into the countryside, spend a weekend camping as far away from street lights as you can, or venture farther north (or south) than you ever have before, to chase the Aurora.

The sky isn't the limit—the sky has no limit.

Sarah Barker

Ideal Observing

Before even taking your first glance up at the sky, there are some simple things you can do to make sure you get the best possible view. Here are a few ways to set yourself up for stargazing success.

Darker is Better

Many of the objects in this book can only be seen at night. When it comes to observing astronomical phenomena, darker is better! When you're trying to stare at distant dots of light, you want to get as far from the distractions of light pollution as possible. Get away from bright lights and glowing cities: the farther you are from city lights, the farther into the heavens you'll be able to see. It doesn't matter how big or powerful your telescope is—if you're trying to use it in Downtown Los Angeles or Central London, you're not going to see much. This is one of the main reasons that professional astronomical observatories are located as far away from major metropolitan areas as they can get. They want to make sure their telescopes are treated to the darkest skies possible.

Get High

Fancy telescopes are also usually found at high altitude—like the observatories on Mauna Kea, Hawaii, La Palma, Canary Islands, or high in Chile's Atacama Desert. This is because the other factor that can wreak havoc on observations is the Earth's atmosphere. You can see this for yourself whenever you see a star twinkle—these are atmospheric effects tampering with your view of the star. Once you're sufficiently high up a mountain, and above more of the atmosphere, you'll get an even better view. Also, since you'll likely be high above the clouds, your chance of getting rained on or clouded out will be significantly lower!

Of course, the best way to overcome atmospheric effects is to make it all the way into space, which is why the Hubble Space Telescope gets all those incredible images (and why astronauts get a pretty good view, too!). However, it is extremely unlikely that we will get the chance to enter orbit, and not many of us can pop over to an astronomical observatory for an evening, either; thankfully there are other ways to set yourself up for success here on Earth …

Choose Your Nights Wisely

Check the weather forecast. If it looks like it is going to be cloudy, you might be better off picking a different night, or a different location. Zero cloud cover is the dream. Also, if you want to look for fainter objects, it is best to pick a night where the Moon is new or doesn't rise until later—unless you want to look at the Moon, of course …

Adjust Your Eyes

Another important trick is to spend some time allowing your eyes to adapt to the darkness. As you've probably experienced when you've stepped into a movie theater, or a dimly lit room on a bright day, it takes a while for our eyes to adjust. The same thing happens if you go outside at night ready to look at the stars. At first, you'll see only the brightest stars, but after a few minutes of uninterrupted darkness, more and more will appear.

It takes a long time for our eyes to fully adjust to a dark sky—somewhere between twenty and sixty minutes! For all that time, you need to commit to the darkness. One flash of light fifteen minutes in and your eyes will be back to square one, which is very annoying, so make sure you turn off your phone! If you find yourself in need of a light, use a red one. Red light doesn't ruin your dark-adaptation, and there are lots of good LED headlamps available with dim, red settings—a sensible purchase for stargazers.

Telescopes
and Binoculars

There are a dizzying number of telescopes out there and choosing the right one for your needs can be frustratingly complicated. Do you want something portable? Affordable? Easy to use? Good for younger stargazers? Powerful? Built for life? It's probably going to be a balancing act between priorities like these, and it may take a bit of research to find the best fit for you. A decent telescope can be a significant investment, and you then have to figure out how to use it,

how to maintain it, where you're going to store it, and so on.

My advice is to leave the telescope until later. Start out with your eyes and catch the stargazing bug. Then search out your local astronomy group and see when they're hosting their next stargazing event. These are a fantastic opportunity to try looking through telescopes with the help of an expert, before committing to buying one yourself.

If you want to see more than is possible with the naked eye, but aren't quite ready to commit to buying a telescope, consider investing in a pair of binoculars. These can be much more user-friendly and easier to set up than a telescope, and a decent pair of astronomy-friendly binoculars can be a more affordable way to take the next step in stargazing. Depending on the pair you choose, you could see as many as fifty times more stars than with the naked eye alone.

When selecting a pair of binoculars you'll notice that the models feature two numbers with an *x* in the middle (8x50, for example). The number on the left is the magnification, (i.e., how powerful the binoculars are), and the number on the right is the aperture, (i.e., the diameter of the large glass lens at the front). You might be tempted to think bigger is better, and go for the highest power possible, but high-magnification binoculars can be pretty big and unwieldy. For a first pair, it might be better to get something smaller and easier to carry around, and that doesn't require a tripod. Binoculars with a maximum magnification of 10x and an aperture of fifty or less should allow you to see many more stars!

Navigating the Skies

It's a big old sky up there, and finding your way around it can take a while. This book contains lots of ways to find different constellations and other fascinating stellar sights, but there are a few bits of background you might find helpful as well.

Know Your Place

The stars look different depending on where you are in the world. The orientation of constellations can change, and entirely different stars can be seen depending on where you're looking from. Things can seem upside down or back to front compared to how they are described in this book, depending on your location. Some of the sights in this book call for you to be at certain latitudes in order to see them—but what does this mean? Latitude is a way to measure your distance from the equator—the imaginary line running around the middle of the Earth, separating the planet into equal northern and southern hemispheres. If you're below the equator, you are in the southern hemisphere, above it and you're in the northern hemisphere. Each hemisphere is divided into 90°, where 0° marks the equator, and 90° the pole. Sydney, Australia, has a latitude of about 34°S, New York City, USA, is about 40°N, London, UK, is farther north still at 51.5°. Sometimes, circles of latitude are called parallels, so, for example, New York would be on the fortieth parallel north.

Rules of Thumb

Another place you might see *degrees* popping up, relates to the sky itself. Astronomers measure the distance between objects in the sky in degrees, similar to the way in which they are used on Earth: 0° corresponds to the horizon and 90° is the point right above your head—the zenith. Halfway between these two points would be 45°. Jumping smaller distances is a little trickier, but luckily, you can use your own hand as a guide. Make a fist and hold it at arm's length, with the back of your hand facing you. The width of your fist is about 10°. Stick your pinky finger up and that covers about 1°. The full Moon is about half a degree, so you should be able to completely cover it with your pinky finger. You can also stretch apart your pinky finger and thumb to make about 25°; your index and pinky fingers to make about 15°; or your three middle fingers to make about 5°. Using some combination of these can help you find your way when guides say things like "Look 15° southwest to see this object …"

Star Charts

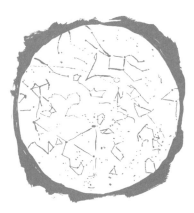

Your view of the night sky changes depending on where and when you are looking, so there is no single map that works for everyone all of the time. Star charts are maps of the night sky for a specific location, on a specific date. You can find these in astronomy magazines, or using free, online resources. Just print out the star chart for the night you want to go stargazing, but make sure you're holding it the right way up. If you're facing due north, hold the map so that the N for north is at the bottom of the page, and the S for south is at the top. If you turn around to face south, rotate the page so the directions are reversed.

Phone Apps

If this sounds a bit confusing, or if you don't want to print a new map every time you look at the sky, fear not— there's an app for that! I find star map apps especially useful because they can react to your position in real time. As you turn your phone from west to east, or tilt it higher in the sky, these apps will refresh their on-screen maps to show what you should be able to see in the direction you're facing. If you're planning a long night of stargazing, make sure you bring a charger. Also, change the settings to red light so your eyes can adapt to darkness.

Extra Tricks and Tools

By now you should have a good idea of where and when to go, and what to take with you. But before you head out looking for things to see in the sky, here are a few additional tips to help maximize your enjoyment.

music and company, and luxuriate in the experience. Stay up late looking for meteors or gazing up at the Milky Way. A few hours spent staring into the heavens can create memories that will stay with you for life.

Join a Club

When you're trying out something new, or have a lot of questions, it can be helpful to find other people with the same interests. Thankfully, there are lots of people interested in astronomy. In fact, there are local astronomy clubs everywhere. Many of the amateur astronomers who make up these communities are happy to bring their telescopes out for a "star party" at a local park or event, and to let you have a look for yourself. Search online for your nearest group and think about attending their next event.

Take Your Time

If you really want to make the most of the sky, get comfy and soak it in. Take a warm blanket, a comfortable camp chair, snacks and drinks, some good

Protect Your Peepers

Never ever, ever stare into the Sun without proper protection. Just looking into the Sun with your naked eye can cause permanent damage to your eyes, while looking through binoculars or a telescope at the Sun can be catastrophic. Sunglasses do not count as protection. If you want to look at the Sun, make sure you have special glasses that are ISO certified, and if you're using a telescope, be sure that you also use the correct solar filter.

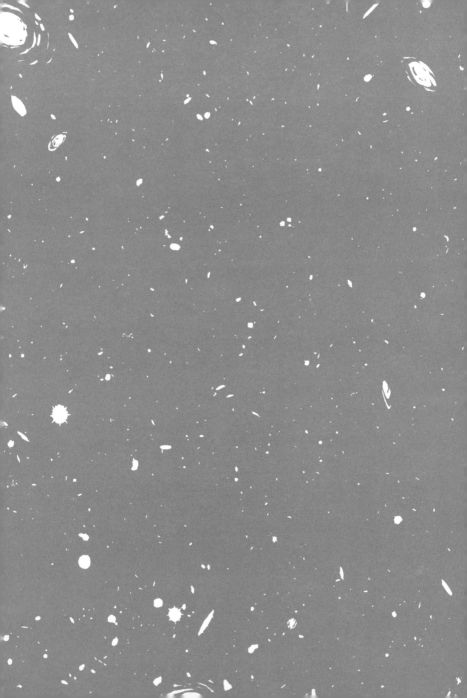

Naked Eye

This section of the book showcases the biggest, brightest and most accessible wonders of the sky. From commonly clocked constellations, to spectacular celestial phenomena, there should be plenty of things here that are ready and waiting for your viewing pleasure. Most of the sights in this chapter can be enjoyed without any special equipment, and are ideal for beginners. A few may require viewers to be in a particular part of the world, or are only observable under certain conditions, but are still fascinating to learn about. Then, hopefully, when you find yourself in the right place at the right time, you will know exactly what those glowing lights in the sky are … (Hint: they're on page 70.)

1

The Milky Way

The Milky Way is a band of stars of our own galaxy beaming radiantly across the night sky. There are at least *one hundred billion* stars in our galaxy, and because we're out in its suburbs, we have the good fortune of being able to see the shape of the Milky Way in our night sky. We're about two-thirds of the way out from the center of the Milky Way, and so when we gaze in that direction at night, we see this beguiling band of stars in our line of sight. The Egyptians called it the Heavenly Nile, and thought it was an extension of their great river into the sky. In Northern Europe it was called the pathway of the birds, as it was believed that migratory birds used the Milky Way to guide them. You can tell where it is using a constellation based on one bird in particular—Cygnus, or, the Swan.

1. In autumn in the northern hemisphere, go somewhere super-dark! The darker it is, the more of the Milky Way you'll be able to see. Choose a moonless night that is free of clouds and try to get as far away from city lights as you can.

DENEB

2. Look for the constellation Cygnus — the Swan. This is a beautifully symmetrical constellation with a wonderfully bright star, Deneb, in the tail of the swan, and the star Sadr in the middle, with the two wings leading off to the sides.

3. Cygnus is flying along the Milky Way! When you look at Cygnus, you're looking directly into the Milky Way, and the gentle glow of billions of stars.

2

Star-Crossed Lovers

Two bright stars in the night sky—Altair and Vega—tell a
story that has been recounted for millennia. In ancient
Chinese legend a humble cowherd (Altair) fell deeply
in love with a princess (Vega). Sadly, their love was forbidden,
and they were forced to live on opposite sides of a great
river (the Milky Way).

The lovers were only able to meet once a year, on the seventh
day of the seventh lunar month, when a flock of magpies would
form a bridge across the river. On this day every year, the legend
is marked by the Qixi Festival in China, and Tanabata—
the Star Festival—in Japan.

At about seventeen light-years from Earth, Altair is one of the
closest stars you can see with the naked eye. Vega is about
twenty-six light-years away, but since it is much bigger and
hotter than Altair, Vega outshines it. In fact, Vega is one of the
brightest stars in the sky. No matter where or when you view
these stars, they will always be on either side of the Milky Way.

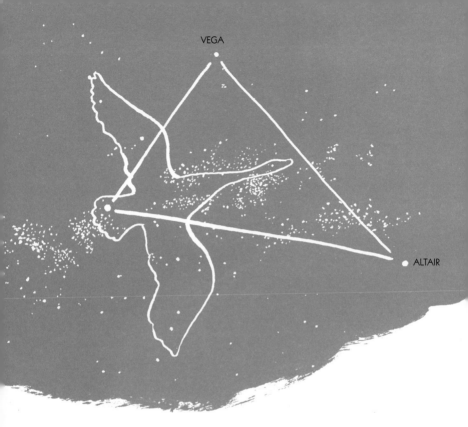

VEGA

ALTAIR

1. Find Cygnus using the steps on page 25.

2. Look for Deneb—the brightest star in the Cygnus constellation—then look for two other bright stars that form a triangle with Deneb. This constellation is called the Summer Triangle.

3. These two stars are Vega and Altair, the star-crossed lovers. If conditions allow you to see the Milky Way, you'll notice that it runs between these two stars, keeping them on opposite sides.

3

The North (for Now) Star
and How to Find It

Faithfully guiding travellers for centuries, Polaris (the North Star
or the Pole Star) might be the most famous star in the northern
hemisphere—but it's not the brightest! (That honor belongs to
Sirius, see page 46.) If you can find Polaris, you can find north.
This is because it lies in a direct line above the north celestial
pole. This is great not only for celestial navigation, but also for
taking long exposure photographs—Polaris creates a seemingly
fixed point in a "starlapse" picture, around which all other stars
appear to rotate.

• TOP FACT! •

The North Star didn't always point north! We're living in a quirky time in Earth's
history where our rotational axis lines up exactly with Polaris's position in the night
sky, but this was not always the case. When the Egyptians were building the
pyramids, a star named Thuban was in pole position instead.

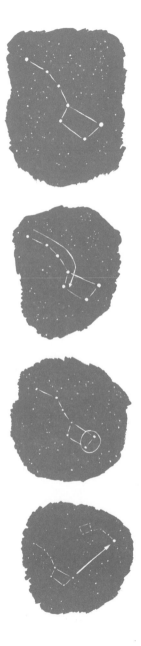

1. Look for constellation Ursa Major, specifically the region known as the Big Dipper, or the Plough. This constellation is made of seven stars and kind of looks like a saucepan …

2. Look at the handle of the "saucepan," and follow an imaginary line from the end of the handle, toward the pan, down the side, then along the bottom of the pan.

3. Locate the far side of the saucepan. The two stars that make up this side of the pan are known as the pointer stars.

4. Draw a line up the side of the saucepan, through both of the pointer stars and keep going in a straight line. The next bright star you hit is the North Star! The pointer stars point to Polaris.

4

The Next North Star

Polaris's reign as the North Star is finite; a successor is waiting in the wings—the star Gamma Cephei (also known as Errai). Due to an effect called the precession of the equinoxes, the north celestial pole changes its position very slightly compared to background stars in the sky. It is a cycle that repeats every twenty-six thousand years or so, and it will be almost two thousand years before Gamma Cephei takes its place in the direction of the north celestial pole. In the meantime, however, this star is worthy of appreciation for two reasons. First, it is a double star, which comprises a bright, Sun-like star, plus a dimmer, red dwarf star orbiting a common center of mass. This is called a binary system. Second, it has a planet! Gamma Cephei was the first binary star system to be found to have a planet, and its giant, orbiting world is almost twice the size of Jupiter.

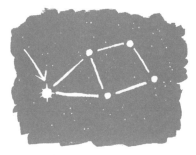

1. Gamma Cephei is in the constellation Cephus, the King. This constellation is circumpolar (visible all year round) for regions far enough north, and visible from as far as 10°S for at least part of the year. It looks a bit like a pentagon, or a simple house shape (with Gamma Cephei as the tip of the roof) but is not as bright as other constellations. To help you locate it …

2. … use the Plough/Big Dipper to find Polaris (see page 29). Hold out a clenched fist at arm's length, placing one side next to Polaris. On the other side of your fist (in the opposite direction to the Plough), you should find Gamma Cephei.

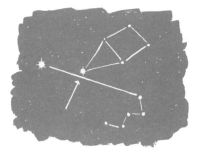

3. Or, find Cassiopeia (see page 50) and make a line from the star at the end of the M (or W) toward Polaris. Just over the halfway point of that imaginary line is Gamma Cephei.

5

Apollo 11 Landing Site

The most accessible wonder of the night sky is our Moon. If
you look through binoculars, or a small telescope, you'll quickly
notice that the Moon is full of intriguing details, but even with
just the naked eye, you can see some fascinating features—
including the area where the first Moon landing took place!
Dark patches on the Moon are called *maria*, or "seas," because
early observers thought that's what they were—dark oceans of
water on the Moon. This is not quite right; they were formed
by ancient volcanic eruptions that spewed basalt rock over the
surface of our dearest natural satellite, but the name remains.
One of the most famous of these *maria* is the Sea of Tranquility.
On July 20, 1969, two astronauts from the Apollo 11 mission
became the first people to walk on the surface of the Moon.
Their landing site was located in the Sea of Tranquility.

TYCHO

1. Pick a night when the Moon is full and orient yourself by looking for one of its most prominent craters—the Tycho crater—in the lower part of the Moon (for observers in the northern hemisphere).

2. Next, look for the dark patch across from the Tycho crater that resembles a lobster claw, or perhaps rabbit ears ... the main part of this is the Sea of Tranquility.

3. Draw an imaginary line from the Tycho crater toward the Sea of Tranquility, and the first dark patch you reach is the site of the Apollo 11 landing—people have walked on that spot!

6

Astronauts!
(The International Space Station)

Did you know that astronauts are orbiting the Earth right now? The International Space Station (ISS) has been continuously inhabited since 2000 and orbits the Earth sixteen times a day! Including its huge solar panels, it's about the size of a soccer field, and orbits at an altitude of about almost 250 miles (400 km) above the Earth's surface. Its crew of between three and six astronauts from different countries are zipping by at a speed of around five miles (8 km) per second, which is nine times faster than the speed of a bullet! And the best part is you can easily see the ISS with the naked eye. In fact, after the Sun and the Moon, it is sometimes the brightest object in the night sky.

1. Find out when the ISS is due to pass over your location. There are many free apps and websites that can figure this out for you, and some can even send you alerts, like www.spotthestation.nasa.gov

2. In the hours immediately after the Sun sets, or just before it rises, the light of the Sun reflects on the solar panels of the space station, but not on the surface of Earth. This provides perfect viewing conditions to catch a glimpse of the ISS.

3. Head outside and try and find an open location away from buildings and trees in order to maximize the amount of sky you can see. The ISS will appear as a very bright star but will be moving rapidly.

4. On a good day, you might be able to watch the ISS for over five minutes as it zooms in an arc above your head. And who knows, maybe the astronauts are looking down at you, too …

7

Shooting Stars

Shooting stars are not, in fact, stars, but meteors—little pieces of space debris burning up in our atmosphere. Some of them are as small as a grain of sand; but although they are small, they are mighty, often creating magnificent streaks of light across the night sky. If the meteor doesn't completely burn up in the atmosphere, and part of it makes it to the ground, then it becomes a meteorite.

Make a wish! For thousands of years, people have believed that shooting stars are lucky—but you don't have to be super lucky to spot one. On an average night, the chances are you'll see one every fifteen minutes, but if you pick a night during a meteor shower, you could see many more. Meteor showers happen when the Earth passes through the ice and dust debris left by a comet (see page 40). They occur multiple times a year and can be witnessed all over the world.

1. Go outside tonight, lie down and get comfortable, wait for about fifteen minutes, and you should see one!

2. To increase your chances, have a look during a meteor shower. Check online for when the next one will take place in your area, or use this rough guide:

LYRIDS—APRIL

PERSEIDS—AUGUST

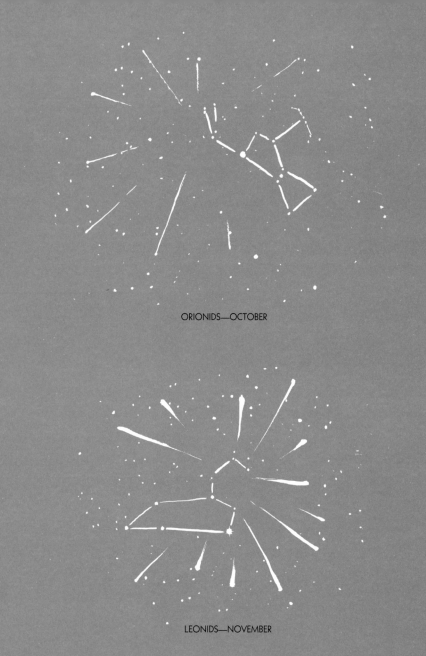

ORIONIDS—OCTOBER

LEONIDS—NOVEMBER

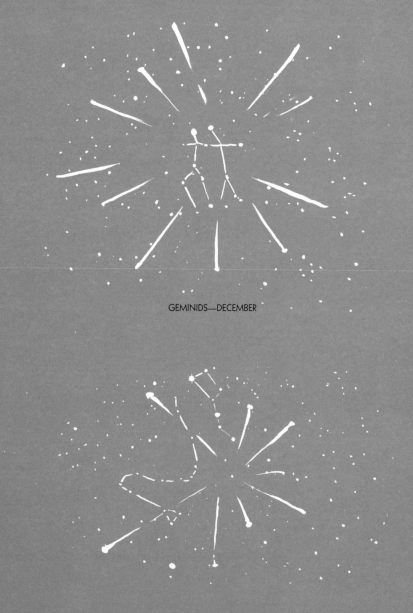

GEMINIDS—DECEMBER

QUADRANTIDS—DECEMBER/JANUARY

39

8

Halley's and Other Comets

Comets are fascinating. Some astronomers believe that they are relics of the formation of the Solar System, and may even be responsible for a portion of the water here on Earth … They are huge masses of dust and ice and spend most of their time on the outer reaches of our Solar System. Occasionally, their highly elliptical orbits bring them into the inner Solar System, where they become visible from Earth—sometimes even with the naked eye. They have two distinct "tails" (one made of dust, the other of ionized gas) that form when the comet gets close enough to the Sun. One of the most famous comets is Halley's Comet, which swings near Earth approximately every seventy-five years. Halley's Comet won't pass again until 2061, but if you can't wait that long, don't worry! You have two options …

1. Every year you can see fragments of Halley's Comet in the form of a meteor shower! Each October, when the Earth passes through the trail of ice and dust that Halley leaves behind, we can see the Orionid meteor shower! As tiny, discarded fragments of Halley's comet burn up in Earth's atmosphere, you can see lots of shooting stars—perhaps as many as fifty an hour.

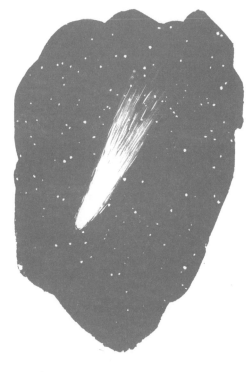

2. Look for another comet! Although super-bright comets are few and far between, there are a handful of comets that can be seen through a small telescope each year. In fact, many comets have been discovered by amateur astronomers and "comet hunters" using small telescopes. Check online for opportunities for comet sightings, and to learn how to hunt for your own …

9

A Very Famous Belt

One of the most striking and identifiable constellations in the night sky is Orion the Hunter. Characterized by a distinctive pattern of stars, Orion can be seen by a huge proportion of the planet. Its recognizable shape and color variation make it an easy one to spot, and this arresting constellation has inspired people across the world for thousands of years. Two bright, giant stars (Betelgeuse and Bellatrix) mark Orion's shoulders, another two (Saiph and Rigel) mark his knees, while three bright stars in a row make up his famous belt. Hanging from this belt you'll see his sword, and in his left arm, you might be able to make out his bow.

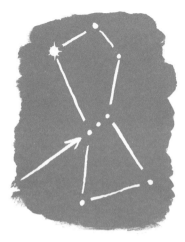

1. The best time to see Orion is between January and March. In the northern hemisphere, look to the south-west in the evening, and try to find the three stars in a line that make up Orion's Belt.

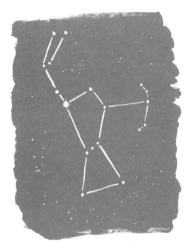

2. From there, you should be able to spot the shoulders, feet and, if it's dark enough, perhaps his bow. In the southern hemisphere, look to the northwest, and if you're close to the equator, Orion will appear in the western sky.

• TOP FACT! •

The more you look at Orion, the more you will discover—especially if you are able to use binoculars or a telescope. Hidden in this constellation are double stars (see page 110), beautiful nebulae (see page 84), and even a supergiant star destined to explode (see page 44).

10

A Red Supergiant (Betelgeuse)

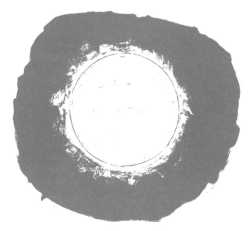

Betelgeuse is a red supergiant star. It is nearly 870 million miles (approximately 1.4 billion km) across—so big, that if placed in the middle of the Solar System it would stretch out beyond the orbit of Mars, and perhaps as far as Jupiter. Compared to Bellatrix—Orion's other shoulder—it appears roughly the same brightness to the naked eye, and since they're in the same constellation, you'd be forgiven for thinking they were near to each other in space. Not so! Betelgeuse is about 640 light-years away, whereas Bellatrix is only 240 light-years away. Betelgeuse is just so luminous, you might be tricked into thinking it's closer. Betelgeuse emits as much light as at least fifty thousand suns and is only a fraction of the age of our home star.

1. Look for the three stars in a line that make Orion's Belt.

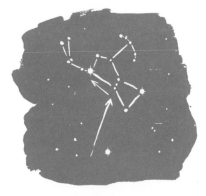

2. Look up and to the left—the brightest star you come to next should appear slightly red.

3. This is Betelgeuse—Orion's right shoulder.

• TOP FACT! •

Within the next million years (quite soon in astronomical terms) Betelgeuse is expected to go supernova. This means it will run out of fuel, collapse under its own weight, and explode spectacularly. It will be so bright that from Earth it will be on show for weeks, if not months.

11

The Dog Star

The Dog Star is a nickname given to Sirius—the brightest star
in the night sky. From the Ancient Greek word for "glowing,"
Sirius forms the heart of the constellation Canis Major, named
by these early astronomers for a faithful hound that followed
Orion the Hunter around the heavens, but other cultures also
associated Sirius with a canine. The ancient Chinese called
Sirius the wolf star, and some Native American tribes also
associate the star with wolves and coyotes. At just under nine
light-years away, Sirius is one of the closest stars to Earth. It
shines with a brilliant white, or blue-white light, but when it is
low in the sky or shining through an unsteady atmosphere, it
can appear to twinkle and shine with different hues.

1. In the northern hemisphere winter, look for Orion the Hunter (see page 42).

2. Follow the three stars of Orion's Belt from his left arm to his right, and continue in that direction.

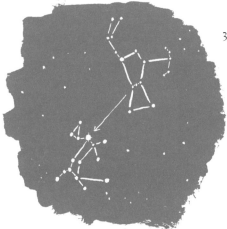

3. The bright star you see is Sirius, the Dog Star!

4. On late summer mornings (in the northern hemisphere), look for Sirius in the east before dawn.

12

A Punished Queen (Cassiopeia)

There is a distinctive constellation that stays visible all year long from locations 34°N or farther from the equator (e.g., New York City, London, Madrid, Tokyo). Comprising five main stars, on first glance it looks a little like a letter M or a W—this is Cassiopeia, the vain queen of Greek mythology. At this latitude, the Cassiopeia constellation is circumpolar, meaning that it never sets below the horizon as it rotates around the north celestial pole. Depending on where and when you view this constellation, it will look more like an M or a W as it makes its way around the pole.

In Greek Mythology, Cassiopeia was an Aethiopian queen who boasted of her unmatched beauty. When she claimed to be more beautiful than the sea nymphs, the sea god Poseidon sent a monster to ravage her country. To appease the monster, the queen offered her daughter, Andromeda, as a sacrifice. Andromeda was rescued by the hero Perseus, and Poseidon punished Cassiopeia by chaining her to a throne in the northern sky forever.

1. Use the Big Dipper to find the North Star (see page 29).

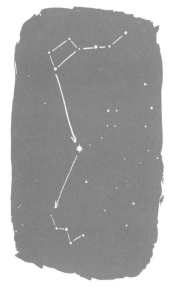

2. From the North Star, look in the opposite direction from the Big Dipper, and you should find the distinctive W or M constellation of Cassiopeia.

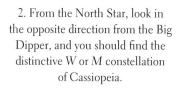

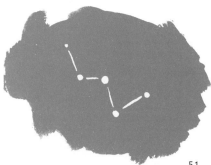

13

The Pleiades

Perhaps the most famous cluster of stars in the night sky,
the Pleiades, or Seven Sisters, has fascinated stargazers since
the dawn of time. Rich in myths and legends, the Pleiades is
actually a cluster of hundreds of stars that all formed from the
same giant cloud of gas around one hundred million years
ago. On a clear enough night in even the most light-polluted
city, you should be able to see a handful of the Pleiades' stars
huddled together. But if you are able to get somewhere dark and
spend some time adapting your eyes to the darkness, you might
be able to make out ten or more stars in this cluster. One of the
best things about the Pleiades is that they can be seen from as
far north as the North Pole, and as far south as Patagonia!
The best time to see them in the northern hemisphere is
in late October and November, when they twinkle almost
from dusk until dawn.

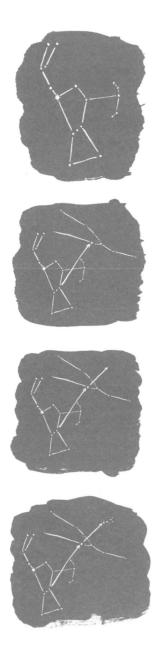

1. Look for the constellation of Orion (see page 42).

2. Follow the path of Orion's Belt from his right side to his left and keep going, past the top of his bow, to the next brightest star in the night sky.

3. Look for a constellation that looks a little like a V. This is part of the constellation of Taurus, the Bull. If the Moon is bright, or you're pretty close to other light sources, you might not be able to see the full V-shape of Taurus's head, but you should be able to see his bright eye, Aldebaran.

4. Look farther to the right, and you should spot a group of stars gathered together—this is the Pleiades!

14

Lunar Halos

Have you ever noticed a large, circular halo around the
Moon? These are caused by tiny, hexagonal ice crystals in the
atmosphere, and are associated with thin cirrus clouds at least
twenty thousand feet (6,000 m) in the sky.

When light from the Moon passes through these tiny ice
crystals they act like little lenses and bend the Moon's light.
Since the crystals are almost identical to one another, they all
bend the light in the same way—typically by about 22°, which
is why lunar halos are also called 22° halos.

If you look closely, you'll see that the inner edge of the halo
is redder than the outer edge—just like a rainbow. The same
phenomenon can occur with the Sun, creating an optical effect
known as a solar halo.

1. Look at the Moon on a relatively clear night, but when there are high, wispy, cirrus clouds in the sky.

2. It's best to pick a night when the Moon is full, or nearly full, so it is as bright as possible.

3. Look for a circle around the Moon. If you hold both of your hands at arm's length and then make them into fists, the width of your fists together is about the size of a lunar halo.

• TOP FACT! •

Every lunar or solar halo you see is unique. Everyone sees their own slightly different halo, as you are seeing light that has traveled through specific ice crystals and into your particular line of sight.

15

The Morning Star

This is the name for a celestial object that is most easily seen
in the hours before dawn and most commonly refers
to something that is not a star, but a planet: Venus.

Venus is the second closest planet to the Sun (after Mercury)
and is usually the closest planet to Earth. You wouldn't want
to go there though. The surface temperature is over 860°F
(460°C), and its atmosphere is made up of carbon dioxide and
sulfur—highly toxic to us humans! Interestingly, Venus rotates
in the opposite direction relative to all other planets in the Solar
System, and very, very slowly. A Venusian day (the time it takes
to complete one rotation on its axis) is 243 Earth days. This
is longer than its year (the time it takes to orbit once
around the Sun)!

1. In the night sky, Venus never strays too far from the Sun and so is best viewed in the eastern sky at dawn, or in the western sky at dusk, when it is referred to as the Evening Star. Which it is depends on its position relative to the Sun: leading the Sun across the sky, it shows up in the morning; trailing, it will linger after sunset. Venus doesn't rise high in the sky but is easy to spot, as it appears brighter than any other planet or star.

• TOP FACT! •

You can tell if you're looking at a planet rather than a star because planets do not twinkle! Stars twinkle because their extreme distance means they appear as a point of light in the night sky small enough for temperature changes in the atmosphere to affect their brightness. Planets are close enough that they appear a little bigger to our eyes and can escape the effect of the atmosphere.

16

A Green Flash

A green flash is a bright peak of green light that can be seen at the very top of the Sun *just* as it sets. It happens when the Sun is almost completely beneath the horizon, with just the tiniest sliver of the top edge still visible. For just a moment, that tiny sliver of sunlight can appear a vivid green.

But what causes this green flash? As sunlight hits the atmosphere, it is refracted (bent) toward the Earth. This means that when we look at the Sun, we're actually seeing an image that is slightly higher than its true position.

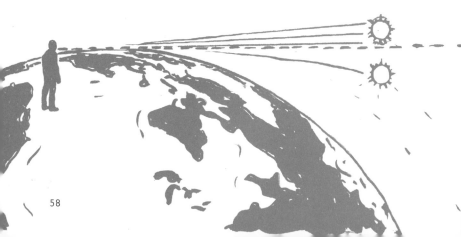

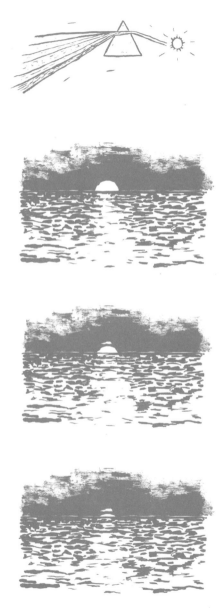

The different component colors of sunlight (see "Blue Skies," page 72) are all refracted by a different amount, with blues refracted more than reds. At sunset, as the Sun starts to touch the horizon, the lower parts of the disc look more red, and the upper parts more yellow. As the lower part sets, we lose the reds, then oranges, then yellows, eventually leaving greens and blues. But, since air molecules scatter blue light away, this briefly leaves one color left to shine—green!

1. Go somewhere with a long, clear view out to the west. Ideally a beach, or even a boat out at sea, where you can see for a long way without obstruction. Choose a day free of clouds to maximize your chances.

2. Wait for the very last moments of sunlight, just before the Sun finally sets, and look toward the sunset. Do you see a small sliver of green?

3. As with any sight related to the Sun—protect your eyes! Do not stare into the Sun. In fact, try not to look at the Sun until the last moment! If you spend too long looking toward the setting Sun, you may impair your light sensitivity and miss the green flash all together!

17

A Solar Eclipse

Total solar eclipses take place when the Moon perfectly passes between the Earth and the Sun, blocking out the Sun. It's as if day becomes night for just a few moments, and only the ghostly tendrils of the Sun's outermost layer (the corona) can be seen. It is a spectacular result of a lucky coincidence: the diameter of the Sun is almost exactly four hundred times bigger than that of the Moon, but the Sun just happens to be about four hundred times farther away. If the Sun were just a little bit bigger, or the Moon just slightly farther away, total solar eclipses wouldn't happen. In fact, the Moon is moving away from Earth at about 1½ inches (4 cm) a year, which means that in a few hundred million years time, the Moon will appear smaller from Earth and total solar eclipses will no longer take place.

1. Find out when the next total solar eclipse is due to take place. A handy feature of this phenomenon is that eclipses are highly predictable. Astronomers can be pretty certain about when and where they are going to occur—you just have to get there (and hope the clouds are kind).

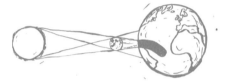

2. Make sure you are in the path of "totality," the area of Earth that will be in the full shadow of the Moon.

3. Wear properly rated safety glasses! Staring at the Sun during an eclipse can cause lifelong damage to your eyes; don't take chances.

4. Wait for the Moon to start blocking your view of the Sun! As soon as the Moon's edge overlaps the edge of the Sun, "first contact" has taken place and the eclipse begins. When the complete disc of the Moon fully blocks the Sun it is called *totality*— the maximum phase of the eclipse. This is when the sky goes dark, the temperature drops, and animals become agitated …

18

Supermoon!

Supermoon is a relatively new term for an astronomical event
that has been taking place for millennia—the appearance of a
particularly large and wonderfully bright Moon. Supermoons occur
when two things happen simultaneously: first, there is a full moon;
second, the Moon is near to its closest distance from Earth. The
Moon's orbit is not a perfect circle, it is a slightly elongated ellipse.
This means that the distance between the Moon and the Earth
changes over the course of the month. On average, the Moon is
approximately 238,000 miles (383,000 km) away from Earth. At its
furthest (the technical term for this is apogee), the Moon is more like
252,000 miles (405,000 km) away, while at its closest (perigee), the
Moon can be as little as 224,000 miles (360,000 km) away. These
orbital changes are due to gravitational interactions with the Earth,
the Sun, and even the other planets in the Solar System.

AT PERIGEE   AT APOGEE   COMPARISON

1. Supermoons take place a couple of times each year.

2. For the most spectacular view, make a note of the times of moonrise and moonset, and be ready to look for the Moon when it is closest to the horizon.

3. If you are *really* good at planning ahead, mark your calendars for November 25, 2034, when you will see the closest supermoon since 1912, and December 6, 2052, for the closest supermoon of the twenty-first century!

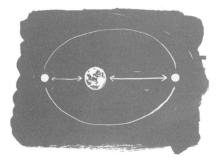

• TOP FACT! •

A supermoon can be up to 30 percent brighter and 14 percent bigger than other full moons. However, this is not actually enough to be noticeable with the naked eye. If you think the Moon looks larger than usual, this may be due to Moon illusion, which is when the Moon's proximity to the horizon tricks your mind into thinking the Moon is particularly big.

19

Noctilucent Clouds

Visible only at night, these are the highest clouds in the atmosphere, and their spectral forms can provide a striking sight on a summer's night. Glow-in-the-dark clouds? Yes, please!

Noctilucent clouds form in the mesosphere, as high as fifty miles (80 km) above the surface of Earth, where temperatures can dip as low as -220°F (-140°C)! Their glowing strands are most likely made up of ice crystals. They're so bright that they've even been witnessed by astronauts on board the International Space Station! There have been reports of particularly bright noctilucent clouds that reflected enough light to allow people to read newspapers at night. These instances may have followed volcanic eruptions and asteroid impacts.

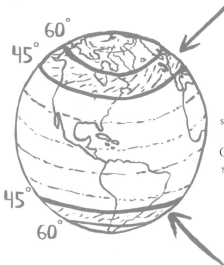

1. Head to high latitudes—between 45° and 60°N or S. This would be somewhere between Madrid in Spain and the Shetland Islands in the UK, Oregon and Alaska in the USA, or the southernmost parts of New Zealand's South Island.

2. For the best chance of success, keep an eye out in the summer—choose nights between May and August in the northern hemisphere, or November to February in the southern hemisphere.

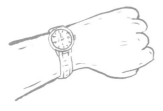

3. Set a reminder to look up about two hours before dawn or after sunset, as these clouds are best spotted when the Sun can't be seen from ground level but is still shining at higher altitudes.

4. Noctilucent clouds are sometimes visible a day or two following a large rocket launch, as the rocket exhaust spews great volumes of water into the upper atmosphere. So keep an eye on when SpaceX or NASA plan to send something particularly big into space!

• TOP FACT! •

Noctilucent clouds are a useful indicator of what's going on in the higher altitudes of the atmosphere and can help scientists study climate change.

20

The Southern Cross

Just as the North Star is visible all year long to those in the far north, the Southern Cross (or Crux) is visible all year long to those in the far south. It is the smallest of the eighty-eight official constellations, but has been guiding navigators for centuries because it can be used to find the south celestial pole. The Southern Cross has major significance for several southern hemisphere nations and appears on the flags of no fewer than six countries. Even those in the southernmost regions of Florida or Hawaii can catch a glimpse for a short time during the year, but to guarantee a good look, head to latitudes at least 35°S. Comprising four stars, the Southern Cross looks a little like a kite to some, as one bar is longer than the other. This longer bar points toward the south, but unlike the northern hemisphere, there is no star sitting directly on the south celestial pole. Instead, you can use two more stars to make sure you're looking in the right spot.

1. Find the Southern Cross. These four bright stars are joined in a recognizable pattern and are actually joined by a fifth star, visible if you look closely.

2. Look for two more bright stars nearby; these are the pointer stars and are in the constellation Centaurus.

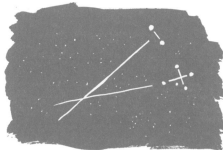

3. Draw a line through the longer arm of the cross, and another from the middle of the pointer stars.

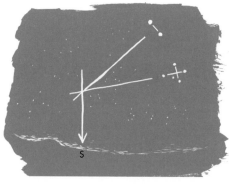

4. The point at which these two lines meet is the south celestial pole! Look down from this spot, and the point at which you reach the horizon is due south.

21

Aurora

These mesmerizing, bewitching bands of color in the night sky enchant all who are lucky enough to see them. In the northern hemisphere they are known as Aurora Borealis, and in the southern hemisphere, Aurora Australis. People travel for thousands of miles to try and catch a glimpse of this natural light show. But what is it, exactly?

It turns out that it's all about magnetism. The Earth and the Sun both have their own magnetic fields. In fact, the Sun has many magnetic fields, and as it rotates on its axis, these fields become tangled together, causing something called sunspots (see page 82). A stream of electrically-charged particles (solar wind) can be propelled from these sunspot regions, reaching speeds of five hundred miles (800 km) per second! When this solar wind reaches the Earth, it is funnelled to the poles by Earth's magnetic field, where it interacts with charged particles in the atmosphere to create the hypnotic light shows of the Aurora!

• TOP FACT! •

Auroras also occur on other planets! Jupiter, Saturn, Uranus, Venus, and Mars are all capable of producing some sort of aurora, and there's no reason to suspect that planets outside our Solar System wouldn't have their own northern or southern lights, too!

1. Head to the Arctic or Antarctic circles during the winter when the nights are long and dark, and look up!

2. To maximize your chances of seeing an aurora, choose a year when the Sun is more actively producing sunspots. The Sun follows an eleven-year pattern of sunspot activity, and the next maximum is predicted to take place in 2024, so plan your trip between 2022 and 2027 for the best results!

22

Blue Skies

Have you ever wondered why the sky is blue? You're not alone. Smart people spent a long time figuring this one out, but now we know!

The daytime sky is lit by light from the Sun. Light from the Sun appears white or colorless to our eyes, but in fact is made up of all the colors of the rainbow! You can test this by using a prism. A prism is a specially-shaped crystal that can separate white light into its component colors by bending (or refracting) each color to a different degree. The different colors correspond to different wavelengths of light; red light has a longer wavelength, while blue light has a shorter wavelength.

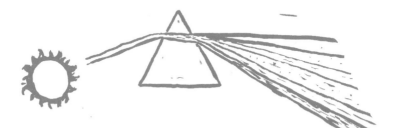

When sunlight reaches Earth's atmosphere, it is scattered by the gases and particles in the air. Due to its shorter wavelength, blue light is scattered much more than other colors, and so we see a blue sky!

You can see this effect even more clearly at sunset. As the sun dips closer to the horizon, its light passes through more and more of the atmosphere in order to reach your eyes. So much atmosphere in fact, that the blue light is scattered altogether before it reaches your eyes, leaving more of the oranges.

1. This one is the easiest in the book to spot! Wait for a cloudless day and look up!

2. Compare the color of blue right above your head to the color close to the horizon; the sky above you should be a slightly darker shade of blue. This is because as the sunlight passes through the air, the blue light has been scattered many times in many directions, and at the same time the surface of the Earth has also reflected and scattered light, further diffusing the color.

23

Sun Dogs

Unfortunately our home star has no canine companions. Sun dogs (or parhelia) is the term given to an atmospheric optical phenomenon similar to a solar or lunar halo (see page 54).

Sun dogs can be a pretty incredible sight, especially very bright ones, as they can create the illusion that there are three suns in the sky! In fact, what we see is light from the Sun passing through ice crystals in the atmosphere, which gather in cirrus clouds or, in extremely cold conditions, fall through the air at low altitude. Either way, the crystals are hexagonal in shape, and will bend light from the Sun to create two reflections of our home star in the sky. The reason they appear on either side of the Sun is that as the crystals fall through the air they become vertically aligned and refract the sunlight horizontally, producing the effect we observe.

Moon dogs can also be seen but are much rarer. This is because the Moon needs to be full, or nearly full, in order to produce the effect, whereas sun dogs can be seen at any time of the month.

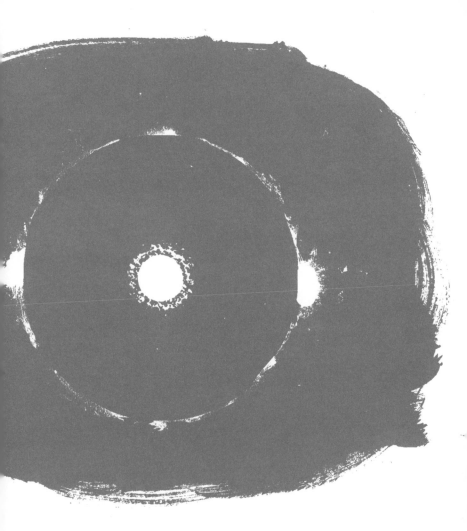

1. Sun dogs can be seen from anywhere in the world at any time of year, but to maximize your chances wait for a clear day when the Sun is close to the horizon (rising or setting). Always take precautions when gazing at or near the Sun, as looking directly into the Sun can cause permanent damage to your eyes.

24

Magellanic Clouds

In southern hemisphere skies, on a dark night, two intriguing
shapes can be seen with the naked eye. They look a little bit
like pieces of the Milky Way that have drifted away from the rest
of the galaxy, but in fact they are completely different galaxies!
They are the Large and Small Magellanic Clouds: two irregular
dwarf galaxies that are companion galaxies to the Milky Way.
For a long time they were thought to be orbiting the Milky
Way, like little galactic pets, but recent observations suggest they
are moving too fast to be in orbit and may eventually collide
with the Milky Way. The Large Magellanic Cloud is around
160,000 light-years away from the Milky Way and contains
around 30 billion stars. The Small Magellanic Cloud is even
farther away, around 200,000 light-years, and contains around
3 billion stars. Compared to the Milky Way's 250 billion (or so)
stars, these are pretty puny galaxies.

1. Head south! They can only be seen below 17°S latitude, and the farther south you go, the better the view.

From 20°S and below, the Magellanic Clouds are visible all year round (on clear nights!).

2. If it's dark enough, the Large and Small Magellanic Clouds should be easy to spot if you look toward the south celestial pole (see page 67). But for some extra help, you can use the bright stars Sirius and Canopus. Draw a line from Sirius (the brightest star in the sky, see page 46) to the next brightest star, Canopus. Keep going in that direction and you'll find the Large Magellanic Cloud.

Farther Afield

This section of the book features objects
and events that are a little trickier to spot than
those you've seen already. They demand a
little extra effort, and perhaps a little more
luck too—you'll want incredibly dark skies, or
ideal viewing conditions. Many of these sights
require the use of a telescope in order to be
truly appreciated, so if you have access to dark
skies and a good telescope, this is the section
for you! If you don't have a telescope or some
binoculars at home, don't worry! Reading about
these fascinating phenomena is still enjoyable.
And if you find yourself with access to a
telescope—at a star party or astronomy club
meeting, for example—you can try looking for
some of these space oddities yourself.

25

Andromeda Galaxy

Did you know that you can see an entire galaxy with your naked eye? The sky does need to be incredibly dark, and your eyes need to be dark-adapted, but it is possible! Andromeda is the closest galaxy to our Milky Way and is perhaps the most distant object observable with the naked eye. But if you have a pair of binoculars or access to a telescope, you'll get an even better view. Andromeda is a beautiful spiral galaxy about 2.5 million light-years from Earth. It is at least as big as our Milky Way, and a 2006 study suggested it contains as many as a trillion stars! Amazingly, it is the only major galaxy that does not appear to be speeding away from us as the Universe expands; instead, it's headed straight for us! Astronomers predict that Andromeda will one day collide with the Milky Way, but the vast distance between their component stars means that of the hundreds of billions of stars on a collision course, only one or two will actually collide.

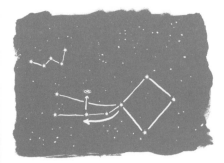

1. Choose a super-dark location, on a super-dark night. Make sure the Moon is not due to rise, and that you are as far away from light pollution as possible. In the northern hemisphere, the best time to look is winter. Spend some time getting your eyes adjusted to the dark (put your phone away!) and then either …

2. … look for the constellation Cassiopeia (see page 50). You will notice that one 'point' of the "W" (or "M") is a little deeper than the other—this is your arrow. Follow it in the direction it points for about three times the size of the Cassiopeia constellation. You'll know you're looking at Andromeda when you see what appears to be a wisp of smoke, frozen in the sky.

3. Alternatively, look for the Great Square of Pegasus—four stars that form a noticeable square shape, below and to the right of Cassiopeia. The upper left corner of this square is the star Alpheratz (Alpha Andromedae), and seems to branch into two paths (almost like a kite with tails). This is the Andromeda constellation. Follow the upper tail for two stars (about halfway), and then look above them to find the Andromeda Galaxy!

26

Sunspots

Sunspots are dark patches on the surface of the Sun (which is also known as the photosphere). These are regions of very strong magnetic field that push up and through the surface. Sometimes there are very few sunspots, while at other times they are seen in abundance. In fact, sunspot activity follows a well-studied pattern. Every eleven years or so, the Sun is at maximum magnetic activity causing a greater number of sunspots, but then the magnetic activity decreases. This is called the solar cycle and is caused by the Sun reversing its magnetic poles every eleven years. Even though sunspots are pretty small relative to the Sun, they are actually as big as fifty thousand miles (80,000 km) in diameter—more than six times the size of Earth! They appear dark because they are much cooler than their surroundings. The temperature of the photosphere is typically about 10,832°F (6,000°C), but sunspots are around 8,132°F (4,500°C). That is a big enough difference to make them appear much darker.

1. Looking at the Sun is dangerous! It can permanently damage your eyes if you're not careful. If you have any doubts about safely observing sunspots, ask for help from a local astronomy group. If you are using a telescope to look at the Sun, be absolutely sure to use a filter designed for solar viewing.

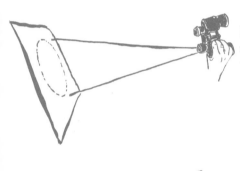

2. One way to see sunspots involves a pair of binoculars and a flat surface like a wall, or a big sheet of paper. *Do not* look through the binoculars at the Sun. Instead, keep the cap on one of the tubes, and point the other at the Sun, with the smaller lens pointed at a flat surface. Attach the binoculars to a tripod if you can, to keep them stable, and shade your flat surface from direct sunlight.

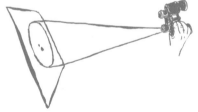

3. Alter the distance between the binoculars and the flat surface so that the disc of the Sun has a distinct edge. You should now be able to make out sunspots! Don't keep your binoculars pointed at the Sun for too long, as the heat could damage them!

4. To maximize your chances, choose a year when the Sun is more actively producing sunspots. The next maximum is predicted to take place in 2024, but even in less active years they are still visible.

27

Orion Nebula

The Great Nebula in Orion is a stellar nursery in the sky. It is where vast numbers of stars are born, and it is spectacularly beautiful. It is an enormous expanse of gas and dust, so huge that it would take light about twenty years to get from one side to the other. At roughly 1,500 light-years away, it is the closest large star formation region to Earth and is bright enough to be seen with a small telescope or binoculars; or, if you are lucky, perhaps even with the naked eye.

The word "nebula" comes from the Latin for "cloud" and is given to night sky objects that appear "foggy" or "fuzzy" due to the amount of gas and dust within them. Some galaxies (e.g., Andromeda, see page 80) were once called nebula, until sufficiently strong telescopes were developed that could show them in more detail.

1. Look for the Orion constellation (see page 42).

2. Cast your eyes below Orion's Belt and to his sword. About halfway down this sword of stars, you should see a faintly fuzzier patch—that is the Orion Nebula!

3. The best time to see the Orion Nebula is when the Orion constellation is high in the sky. In the northern hemisphere, look due south around midnight in mid to late December.

4. The four brightest stars within the nebula can be seen with amateur telescopes and are referred to as the Trapezium.

28

Lunar Craters

One of the most spectacular sights you'll see through a telescope is one of the easiest to make sense of: the Moon. It doesn't matter if you live deep in the countryside, or in the middle of the busiest and brightest city in the world, the Moon is yours to gaze at. It is beguiling through binoculars, or even with no kit at all—but to be really blown away, look at the Moon through a telescope. You'll see mountains, shadows, and crater after crater.

Lunar craters are scars left over from when asteroids, comets, and meteorites smashed into the Moon. Thousands of these mark the lunar surface. Usually circular in shape, some are too small to see from Earth; while others are over 621 miles (1,000 km) in diameter. The reason there are so many is that, unlike Earth, the Moon lacks an atmosphere within which these impactors would burn up. Also, on Earth, plants, animals, and geological activity can hide or remove craters.

1. You can see a few craters with your naked eye; just look at the Moon at twilight, when it is full (or close to full), and look for small, circular shapes with long lines (rays) extending out from them in many directions. The most easily visible crater is Tycho, on the lower half of the Moon (as viewed from the northern hemisphere).

TYCHO

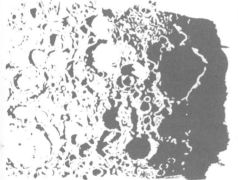

2. For a really special view, get yourself behind a telescope. For moongazing with a telescope, it's actually better to avoid the full Moon and choose a night when the Moon is about half-full instead (the technical term is *first quarter* or *last quarter* depending on whether the moon is waxing or waning). The Moon can be dazzlingly bright when full, and you can make out more detail when you don't have to contend with the brightness. Pan the telescope over the surface of the Moon and take it all in!

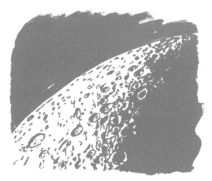

3. Try looking along the terminator—the line that separates the light area of the Moon from the dark. Here you should be able to see long shadows and fascinating detail. You can hop from crater to crater by following this line.

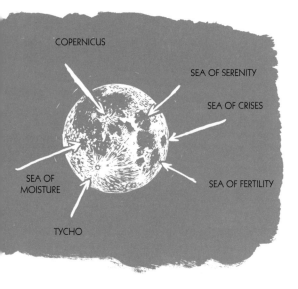

COPERNICUS

SEA OF SERENITY

SEA OF CRISES

SEA OF FERTILITY

SEA OF MOISTURE

TYCHO

4. It's handy to have a map of where you're going, like any explorer would. Search online for "field map of the Moon," and maybe invest in a physical copy if you're serious. With a strong enough telescope you'll be able to see craters, mountains, and even "seas." Enjoy! Be aware that some telescopes can flip images horizontally, or vertically (or both!), so make sure you know what your telescope does before getting confused by a map!

29

Galilean Moons

The Galilean moons are among the largest moons in our Solar System, and fittingly they orbit the largest planet—Jupiter. Their collective name is derived from the man who discovered them—the great astronomer Galileo (1564–1642), who first recorded seeing the quartet in 1610. At first he thought they were dim stars in the direction of Jupiter, but after watching them night after night, Galileo realized that they were orbiting the planet, and must therefore be moons. Back then, only Earth was known to have a moon, and this discovery was one of many to get him in deep trouble with the Catholic church. They eventually accused him of heresy, and put him under house arrest until his death.

The four moons are Io, Europa, Ganymede, and Calisto, and each is fascinating in its own right:

Io is about the same size as Earth's moon and is the most volcanically active object in our Solar System. It spits plumes of sulfur over two hundred miles (almost 350 km) into space and has lakes of liquid lava on its surface.

Europa is one of the most exciting places in the Solar System, because it is a top candidate for extraterrestrial life! Europa is covered by a layer of ice, but below the surface it is believed that there is an ocean of liquid water—where potentially there could be a world of aquatic lifeforms lurking in the dark.

Ganymede is the largest moon in the Solar System and is bigger than the planet Mercury. It is so big, it generates its own magnetic field!

Calisto is one of the most heavily cratered satellites in the Solar System. Its craters are believed to be scars from the initial impacts that shaped its surface over four billion years ago.

1. Find Jupiter (see page 94). With a good pair of binoculars, or a telescope, look carefully to either side of Jupiter. You should be able to make out a line of three or four stars—these are the Galilean moons.

2. If you can only see three, it's because one is behind Jupiter. Check again in a few hours or the next day, and you should be able to see all four.

30

Saturn's "Ears"

When, in 1610, the great astronomer Galileo stared through his invention—the telescope—and set eyes on Saturn, he was the first person to see this beautiful planet as more than a yellow dot in the night sky. But what he saw confused him—it looked as if Saturn was made of three interconnected discs: one in the center flanked by two "ears." Even more peculiar was when he looked back after two years, the "ears" had disappeared!

Over forty years later an astronomer named Christiaan Huygens studied the "ears" in earnest. He figured out they are actually an extensive (but thin) ring of material surrounding the planet. With a fairly basic telescope, you can do the same thing!

1. Saturn is visible for at least part of the year every year. The only times Saturn can't be seen is when its orbit takes it too close to the Sun, as seen from Earth (or when conditions aren't right where you are).

2. It can be easier to spot Saturn when it is in *opposition*, that is, when the Earth comes between the Sun and Saturn, forming a straight line with Earth in the middle. On these dates Saturn will be at its closest approach to Earth, and at its brightest:

- July 20, 2020
- August 2, 2021
- August 14, 2022
- August 27, 2023
- September 9, 2024

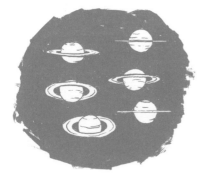

3. Thanks to changes in the orbits of our two planets (Earth takes one year to orbit the Sun but Saturn takes over twenty-nine), Saturn's rings go from almost invisible to fully visible and back again when viewed from earth. This cycle takes about fifteen years.

31

A Storm as Big as the Earth

Jupiter is a planet that holds many records; it's the biggest planet in the Solar System, with the shortest day, the largest atmosphere—and the Universe's most famous red spot. Easily recognizable, just south of the planet's equator, Jupiter's red spot is actually a giant, swirling storm, with winds that can blow over 375 miles per hour (600 km/h). Unlike storms here on Earth that last a few days, there are records of this red-spot storm from as far back as the 1600s, and it could be even older! But what it lacks in youth, it makes up for in size. This spot is a true monster. It's so big, in fact, that the entire Earth could easily fit inside. When it was first discovered, the storm was much bigger—perhaps four times larger than the Earth. But over the intervening centuries it has gradually shrunk and some scientists think that its days are numbered.

1. Find Jupiter.
It is one of the brightest objects in the night sky, and unlike stars it doesn't twinkle. However, like all the planets, Jupiter's position changes in the sky throughout the year, and so it's a good idea to check astronomy catalogues, go online, or use an app to find out when is the best time to see it from your location.

2. To maximize your chances of spotting it (pun intended), choose a time when the red spot will be near the center of the planet—again there are websites and apps for this. Since Jupiter rotates so rapidly, these ideal viewing times will only last a couple of hours each time.

3. If you have the option of using a telescope with a light blue or light green filter, this will help, too.

32

Martian Ice Caps

Mars, the fourth planet from the Sun, was named after the Roman god of war. Covered with a surface of iron oxide that gives the planet its signature red color, Mars has captivated the imaginations of people throughout history, and is frequently touted as the next target for human exploration.

One thing you might not know about the red planet is that it has frozen, permanent ice caps at both its north and south poles. During the winter at each pole, temperatures dip as low as -238°F (-150°C), causing some of the surrounding atmosphere to freeze solid! However, since the martian atmosphere is primarily made of carbon dioxide (CO_2), this means that the poles are coated in frozen CO_2—also known as dry ice.

Getting a good picture of the martian ice caps themselves is a real challenge for an amateur observer. You have to have the best kit, the best atmospheric conditions, and the best luck to have any hope of capturing a decent image. But, catching a glimpse of the planet is much easier!

1. The best time to see Mars depends on where Earth is along its orbit around the Sun, and also where Mars is relative to Earth. More than any of the bright planets, the appearance of Mars changes from year to year. This is because Mars is a relatively small planet, and so when the orbits of Mars and Earth take us far apart, Mars can be much trickier to spot than when our orbits bring us closer together.

2. To know when the next best time to look for the red planet will be, check online, or contact your local amateur astronomy group.

3. You'll know you're looking at Mars because even with the naked eye it glows an orange-red color—hence the name "red planet." Also, since Mars is a planet, it won't twinkle like the stars do.

4. If you're looking through a telescope and want to try and spot a martian ice cap, the northern ice cap is twice as big as its southern counterpart and should be easier to see.

33

Swan Nebula

Also known as the Omega Nebula, Horseshoe Nebula, and M17, the Swan Nebula is an area of star formation about five thousand light-years from Earth. It is one of the largest star-forming regions in our galaxy. The various names have been derived from different parts of the nebula, and the delicate, wispy curve of the swan's head is one of the most difficult shapes to make out—but keep trying! A mass of gas and dust, the Swan Nebula is thousands of times bigger than our entire Solar System! The dust is so thick that many stars are hidden from view, but the nebula glows with the light from hundreds of young stars. In fact, the Swan Nebula holds one of the youngest star clusters in the galaxy, at just one million years old. This nebula is a great one to try and see with a telescope, or even binoculars, as it is one of the brightest in the night sky.

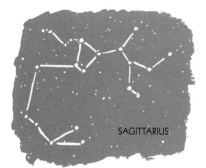

1. The Swan Nebula, like a number of deep-sky objects, can be found using the constellation of Sagittarius. Look for Sagittarius the Archer near to the center of the Milky Way.

2. Hidden in the Sagittarius constellation is a pattern of stars (or asterism) known as the Teapot. Look for this pattern of stars, and then draw a line from the bottom right of the Teapot (i.e. the star Kaus Australis), through the opening of the lid (the star Kaus Media), and continue up about a fist's length until you hit a fuzzy blob in the sky. This fuzzy blob is the Swan Nebula, although there is another fuzzy blob very close to it, the Eagle Nebula (see "Pillars of Creation," page 136).

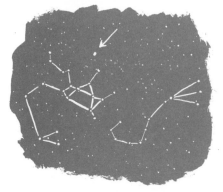

3. The Swan Nebula is best viewed on clear nights in August in the northern hemisphere, when Sagittarius, the Teapot, and the Swan Nebula can all be seen high in the night sky toward the south.

34

A Bubble in Space

It turns out the Universe can blow bubbles. In the constellation
Cassiopeia, there is a super-hot, super-massive star blowing a
giant bubble of gas into space. This object is known as NGC
7635 or the Bubble Nebula. It is just over seven thousand light-
years away from Earth and around seven light-years across. The
young star at the center of the Bubble Nebula is forty-five times
more massive than our Sun, and gas on its outer edges gets
so hot it escapes into space at speeds of over four million miles
per hour (6.4 million km/h)! This fast-flowing gas from the outer
atmosphere of the star sweeps up cooler gas in its path, forming
the outer edge of the bubble. Radiation from the star nearest its
center heats up the gas of the bubble, causing it to glow.

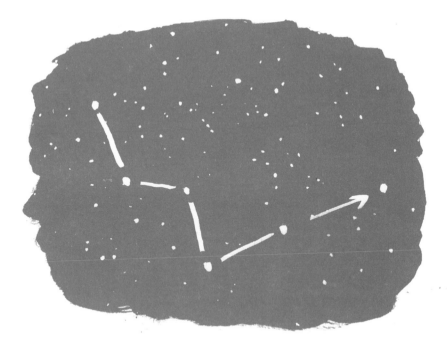

1. This is not an easy one for small telescopes, as the bubble is fairly faint, and diffuse. If you don't have your own six-inch (150 mm) telescope, contact a local astronomy group, which will likely have one that is large enough.
There are also amazing photos online thanks to the Hubble Space Telescope.

2. If you have access to the right kit, and want to give it a try, the Bubble Nebula is best seen from northern latitudes in late summer/early autumn. In August, September, and October it

appears high in the sky, and if you're above 28°N latitude, it stays out all night!

3. Look for the constellation Cassiopeia (see page 50). Trace the W shape, and you will notice that one of V-shapes is a little deeper than the other. Trace along the outer edge of the deeper V and keep going along that imaginary line for the same distance again. You'll know you're at the Bubble when you spot a faint shell of light surrounding a central star.

35

Perseus Double Star Cluster

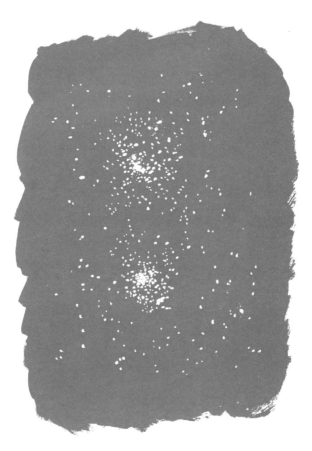

A star cluster is a collection of stars that share a common origin and are gravitationally bound together; these appear as a large group when observed from Earth, such as the Pleiades (see page 52). But what's better than a cluster of stars? Two clusters of stars, right next to one another! In the constellation Perseus, two separate star clusters can be seen side by side. Each contains thousands of stars, some of them hot, supergiant stars that are many times bigger and brighter than our own Sun. The two clusters, named H Persei and Chi Persei, are over seven thousand light-years from Earth, and a few hundred light-years from each other.

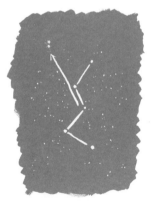

1. The best time and place to spot this double cluster is during winter in the northern hemisphere. It is just about visible with the naked eye if conditions are perfect and looks great through a pair of binoculars or a small telescope.

2. On a dark and clear night, look for Cassiopeia (see page 50) when she is high in the sky. The Perseus Double Cluster lies between this constellation, and neighboring Perseus.

3. Draw a line from Cassiopeia's central star (Gamma Cassiopeiae), through the star in the middle of the shallower V of Cassiopia's signature W-shape (Delta Cassiopeiae) and keep going for about three times the Gamma–Delta distance. This is where you should find the Perseus Double Cluster.

• TOP FACT! •
The Perseus Double Cluster resides in a different arm of the Milky Way galaxy than does Earth.

36

"Little Green Men" (Pulsars)

Over fifty years ago, an astronomer named Jocelyn Bell (now Dame Jocelyn Bell Burnell) made the first observation of a baffling phenomenon. Using a radio telescope, she observed a distant object that seemed to be flickering or pulsing about once every second. It was a very regular pattern in a specific radio frequency that repeated for days. Could it be a signal coming from another civilization? Were "little green men" trying to make contact? Her team genuinely considered this scenario, but when Jocelyn found another pulsating radio source in a completely different place, they realized they hadn't been intercepting alien messages, but instead had discovered a new type of astronomical object: pulsars.

Pulsars are rapidly rotating neutron stars. They are only about the size of a city, but are as massive as the Sun, meaning they are incredibly dense!

Pulsars form when a massive star exhausts its fuel supply, blasts apart in a supernova explosion, and leaves behind a super-dense core of tightly-packed neutrons.

The pulse comes from each pulsar's extremely powerful magnetic field, which emits strong radio waves from its north and south magnetic poles. If the pulsar's poles are aligned with Earth, then we can see these radio waves every time the pole rotates through our line of sight. It's a similar effect to a lighthouse—as the lighthouse rotates, its light appears to blink on and off as it passes you. The same thing happens with pulsars!

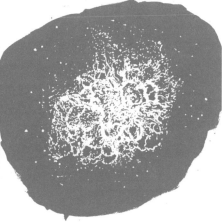

1. Pulsars are far too small to be seen, and mostly they emit light in the radio portion of the electromagnetic spectrum, which our eyes cannot see. They can only be detected in huge, professional radio telescopes. However, there is a famous (observable!) nebula that holds a pulsar in its center: the Crab Nebula.

2. The Crab Nebula can be seen as a fuzzy blob in the night sky through binoculars or a small telescope. With a larger telescope, it appears as a fascinating and complicated cloud of gas and dust, full of intertwined tendrils. It is actually the remains of a giant supernova that was witnessed by astronomers in the year 1054!

• TOP FACT! •

Pulsars are so regular, that they can be used to tell the time! Their pulses keep time more accurately than atomic clocks.

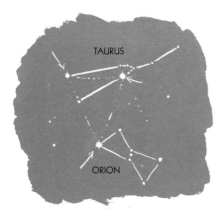

3. On a clear dark night between late autumn and early spring, look to the famous Orion constellation, and the star Betelgeuse (see page 44). Then, find the Taurus constellation, and the bright star Aldebaran (see page 53). Betelgeuse and Aldebaran can form a triangle with another bright star—Beta Tauri.

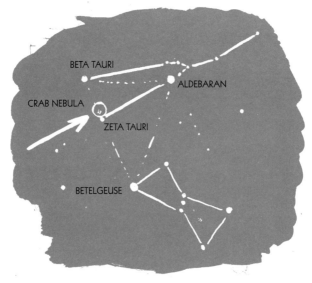

4. Track back toward Betelgeuse from Beta Tauri a little, and you'll find a fainter star—Zeta Tauri. Carefully look around the area surrounding this star, and you should find a small, faint smudge in the sky—that's the Crab Nebula! Remember, at its center is a rapidly rotating pulsar.

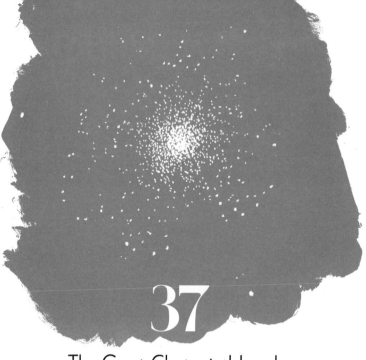

37

The Great Cluster in Hercules

Dotted around the outer regions of the Milky Way (and many other galaxies), are enormous clusters of ancient stars. Known as globular clusters, they can contain hundreds of thousands of stars that are tightly bound by gravity. Globular clusters are some of the oldest objects in the sky, having formed billions of years ago when our galaxy, and even the entire Universe, was much younger. One of the brightest and most famous globular clusters in the northern hemisphere sky is M13, or the Great Cluster in Hercules. This glimmering globe of stars is about twenty-five thousand light-years from Earth and lies outside the disc of the Milky Way. Like many globular clusters, it orbits the galaxy from thousands of light-years away. It was discovered in 1714 by the famous astronomer Edmond Halley (of comet fame, see page 40), who described it rather poetically as "but a little patch, but it shows itself to the naked eye, when the sky is serene, and the Moon absent." If you can get hold of a telescope, you'll get a *much* better view.

1. On a dark, cloudless and moonless summer night (in the northern hemisphere), find the biggest telescope you can and take it outside. A telescope with an eight- to ten-inch (200–250 mm) aperture would be a good choice for this particular object. If you don't own something of this size yourself, track down a star party where people with larger telescopes will be happy to let you take a look.

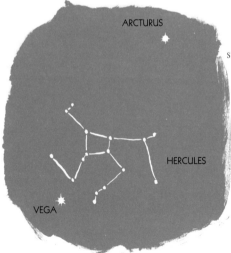

2. Look for the brightest stars in the summer night sky, Vega and Arcturus. Between them, you should be able to make out the Hercules constellation—named after the Roman mythological hero.

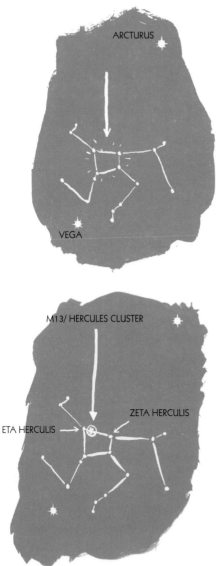

3. Find the four bright stars that mark Hercules' torso. These are often referred to as the Keystone (or less dramatically, the Flowerpot).

4. The Great Cluster in Hercules is found between the two western stars of the Keystone. Draw a line between the star Eta to the star Zeta, and about one-third of the way along you should find the distinctive fuzzy patch of a globular cluster. Through a strong enough telescope, this fuzzy patch will become hundreds of ancient stars gathered together.

38

A Colorful Double (Binary Stars)

Gamma Cephei (see page 30) is a double star. This means that although it might look to the naked eye like one star, it is actually two stars that appear very close together in the sky. And it is not alone. There are many double stars visible from Earth, and some of the most interesting to look at through a telescope are those that shine with different colors, for example, the Albireo binary star system. Albireo is pretty easy to find as it is the head of Cygnus the Swan (see page 24), and bright enough to be seen with the naked eye. However, when you look at this star through a telescope, you get a lovely surprise. Instead of one bright star, you can see two—one glowing with a golden light, the other a bright blue! Albireo is a "true" binary system, meaning that the stars are gravitationally bound to one another, rather than just appearing close to each other by chance.

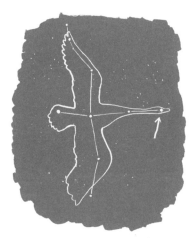

1. To find Albireo, look for Cygnus. Follow the long neck of the swan to find Albireo marking its head.

2. Look at Albireo through a telescope—any telescope will do, but low magnification is better in this case as allowing the stars to appear close together helps the colors to stand out more.

3. Do you see two stars? The star that appears more golden is Albireo A, an orange giant star about five times more massive than our Sun; while the bluer star is Albireo B, a hotter, smaller star about three times more massive than the Sun.

Far, Far Away

This final section of the book focuses on things
that are very far away, or very small, or both.
You're unlikely to be able to see these objects
from your backyard, and in fact, most require
professional equipment to properly observe. But
don't let that put you off! These are perhaps the
most interesting and awe-inducing phenomena
in the book (and in the sky!) and should
hopefully be some of the most fascinating
to read about. Also—thanks to professional
telescopes and spacecraft — if you'd like to see
what these objects look like for real, there are
incredible images for all these topics online.

39

Olympus Mons

Olympus Mons is the biggest volcano in the Solar System. Named after Mount Olympus (the highest mountain in Greece and mythical home of the Ancient Greek gods), Olympus Mons is found near the equator of Mars. It is over three times higher than Earth's tallest peak—Mount Everest. The base is about the size of the entire state of Arizona and almost as big as Italy, and climbs sixteen miles (26 km) into the martian sky. To put this in perspective, the tallest volcano on Earth (Hawaii's Mauna Loa) rises just over 6⅓ miles (10 km) from the ocean floor, but only 2½ miles (4 km) or so are above sea level. Olympus Mons is roughly one hundred times bigger than Mauna Loa by volume. It is a shield volcano—like the volcanoes of Hawaii or the Canary Islands. Rather than erupting violently and spewing molten rock into the sky, shield volcanoes are created when lava flows down the sides of the volcano and travels a longer distance before solidifying. This gives the volcanoes their signature "shield" shape—a gentler slope compared to cone volcanoes.

1. As enormous as this towering volcano is, it's not quite within the reach of a small telescope here on Earth. Some persevering amateur astronomers have used small telescopes to glimpse the ice and clouds that sometimes surround Olympus Mons, but the volcano itself remains elusive.

2. To get a good look at this beast, look for images taken by spacecraft that have been sent to Mars. The Viking 1 orbiter took some fantastic images of Olympus Mons back in 1978, and Mars Express snapped some even better ones in 2004. The Mars Reconnaissance Orbiter got close enough in 2010 to see cliffs along the side of Olympus Mons, possibly formed by landslides.

40

"Failed Stars"

There is a whole class of celestial objects that are not quite
planets, but neither are they stars. With masses of around
fifteen to eighty times that of Jupiter, but only around one-tenth
the mass of the Sun, they fall into their own category—brown
dwarfs. These are made up of the same materials as a star
(mostly hydrogen and helium) but aren't massive enough
to kickstart the process of nuclear fusion in their core. This
means they can't radiate starlight and therefore are often
referred to as "failed stars."

Just like planets, brown dwarfs can have their own atmospheres,
clouds, storms—even aurora—but just like stars, they can have
their own planets. Because brown dwarfs emit so little energy,
they can be tricky to detect. In fact, it wasn't until 1995 that the
first brown dwarf was observed by astronomers.

1. As these little tricksters are so challenging to observe, you'll need to use a properly big, professional telescope to catch a glimpse. Brown dwarfs emit light in the infrared portion of the spectrum, which means our eyes can't see them, so you will need a detector that is designed for infrared. Something like the Keck II telescope in Hawaii, or the Very Large Telescope (honestly, that's its name) in Chile are perfect options.

2. Alternatively, there are spacecraft that were designed especially for this job. For example, the Wide Field Infrared Survey Explorer (WISE), which was launched in 2010 and found several brown dwarfs.

3. You can help discover more! The Backyard Worlds citizen science project allows people with computers to click through WISE images and search for new objects like brown dwarfs.

41

Enceladus' Ice Volcano

Enceladus is not the largest moon in our Solar System (its radius is only 157 miles [252 km]). It's not even Saturn's biggest (there are five greater in size)—but it is definitely one of the most exciting. Covered in ice, and highly reflective, Enceladus is the brightest world in our Solar System. But that's not all, this moon is home to something truly spectacular: ice volcanoes. Hidden beneath its icy crust lies a global ocean of liquid, salty water. Powerful hydrothermal vents shoot jets of icy particles from this ocean out into space. These ice volcanoes eject material at speeds of roughly eight hundred miles per hour (400 m/sec), and this constant stream of icy particles helps feed one of Saturn's rings! With an ocean of liquid water and such fascinating hydrothermal activity, Enceladus is a prime target for the search for life in our Solar System.

1. Although Enceladus can be seen through powerful telescopes as a small dot orbiting Saturn, the best views of this intriguing ice world come from the Cassini spacecraft.

2. Cassini launched in 1997, arrived at Saturn in 2004, and spent the next thirteen years taking jaw-dropping images, and collecting valuable scientific data from Saturn and its moons and rings.

3. The Cassini spacecraft not only discovered the ice volcanoes of Enceladus, but passed directly through the plumes during one of many fly-bys of the moon.

42

Hot Jupiters

The eight planets that make up our Solar System are not the only planets out there. In the 1990s, astronomers discovered planets orbiting stars other than our Sun. Since then the field has boomed and thousands of planets have now been found orbiting other stars. These planets are called extrasolar planets or exoplanets for short. One of the most fascinating things about exoplanets is that the first few to be discovered were nothing like those found in our own Solar System and caused some head-scratching among the experts. On the one hand they were massive gas giant planets (like Saturn or Jupiter), but on the other they orbited incredibly close to their host stars (closer than Mercury orbits the Sun). This earned them the name "hot Jupiters," and to this day astronomers aren't entirely sure how they came about. Did they form farther out and then migrate in? Or have they always been searingly close to their stars? The debate continues …

1. It's pretty tricky to spot an exoplanet, which explains why they were only discovered relatively recently. Planets are so much dimmer than the stars they orbit that any image we might hope to see of them could easily be "drowned out" by that of their parent star. Luckily, there are now several clever methods for detecting them, using professional telescopes on Earth and in space.

2. Look for the wobble. It's not completely true that planets orbit stars—although stars have their planets in tight gravitational grips, planets also exert a small tug on their stars. Both objects actually orbit around a common center of mass, and although this is vastly closer to the star than the planet, the slight movement of the star as it orbits this point can create a detectable "wobble."

3. Watch for transits. Take multiple readings of the light level from a star over time, and if these readings periodically dip, it could be because there is a planet passing between the star and Earth—an exoplanet!

43

A Baby Solar System

The closest region of star formation to Earth is roughly 450 light-years away, in the constellation of Taurus. In this enormous cloud of dust and gas lies a stellar nursery where new stars are forming or have recently formed. One of these stars, named HL Tauri (HL Tau for short), is just one million years old — that may sound like a lot, but compared to our Sun's age of 4.6 *billion* years old, this star is just a baby. Since the region where HL Tau is located is so packed with gas and dust, it's very difficult for telescopes like the Hubble Space Telescope to get a good look at the infant stars within.

However, there are telescopes that can look in other wavelengths of light to the visible light that our eyes (and the Hubble Space Telescope) can see. Telescopes sensitive to the radio and infrared parts of the electromagnetic spectrum can detect the glow that comes from the warm dust and gas of these new stars.

The Atacama Large Millimeter/sub-millimeter Array (ALMA) telescopes in Chile can do just this, and in 2014 they trained their sights on HL Tau, revealing something truly spectacular. The young star was caught in the act of planetary formation! In extraordinary detail, the ALMA images show bright, concentric rings of material encircling the star. Over time, this material will gradually coalesce into planets, comets, and asteroids. In fact, the dark spaces separating the rings in the ALMA image indicate where the beginnings of planets (sometimes called planetesimals or proto-planets) are forming right now. HL Tau has its own baby solar system and gives an indication of what our own Solar System might have looked like billions of years ago.

1. Head to Chile's Atacama Desert, and to the ALMA Observatory. Located 16,400 feet (5,000 m) above sea level, this is a small army of sixty-six radio telescopes, each with a diameter of between twenty-three and just over thirty-nine feet (7–12m).

2. Separate several of these telescopes by distances of up to 9⅓ miles (15 km) — should be no problem! — then point them deep into the Taurus constellation and at HL Tau.

3. Admire the incredible images of a baby solar system!

Spirals and Ellipticals

Galaxies are some of the biggest objects in the Universe. They are like enormous stellar cities, each containing millions, if not billions of stars, as well as planets, moons, comets, asteroids, dust, and gas—even black holes—all bound together under gravity. And yet just like us, galaxies come in different shapes and sizes.

The American astronomer Edwin Hubble (the famous space telescope is named after him) spent years studying galaxies and devised a way to classify them depending on their shape. Although his categories are a little outdated now, the general principle remains: galaxies come in spiral, barred spiral, elliptical, or irregular shapes.

With a powerful enough telescope, you can see examples of all of these. However, the telescope would need to be pretty powerful! It's also important to have a truly dark sky, as far away from city lights as possible. Be aware that most of the detailed images of galaxies are taken with long exposures or use multiple exposures. Even through a powerful telescope, they can look like fuzzy blobs or clouds to our eyes.

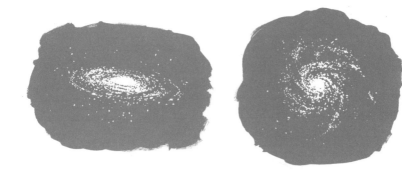

Spirals

These are some of the most beautiful galaxies. With their characteristic swirling arms, and bright, shining nucleus sitting in the center of a flattened disc, they are an enchanting sight. Surrounding the flattened disc of a spiral galaxy is a halo containing clusters of old stars.

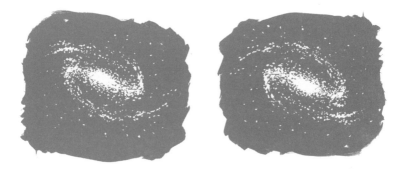

Barred Spirals

These are just like spiral galaxies, except the arms curl out from a bar, rather than a central point. Between about one-third and a half of spiral galaxies have these central bars and our home galaxy, the Milky Way, is one of them.

Ellipticals

These galaxies lack some of the style and grace of their spiral
counterparts but are fascinating nonetheless. They are shaped
like stretched out spheres appearing as ovals, or ellipses in
the sky, with a flat, bright light that dims toward the edges.
Ellipticals are usually quite old, and thought to be remnants
of mergers between smaller, equally sized galaxies.

Irregular

These are the oddballs. Not spirals, not ellipticals, they are often
completely muddled in form, as if perhaps they'd survived a
crash with another galaxy. Our Milky Way has two companion
irregular dwarf galaxies called the Magellanic Clouds
(see page 76).

45

A Light Echo

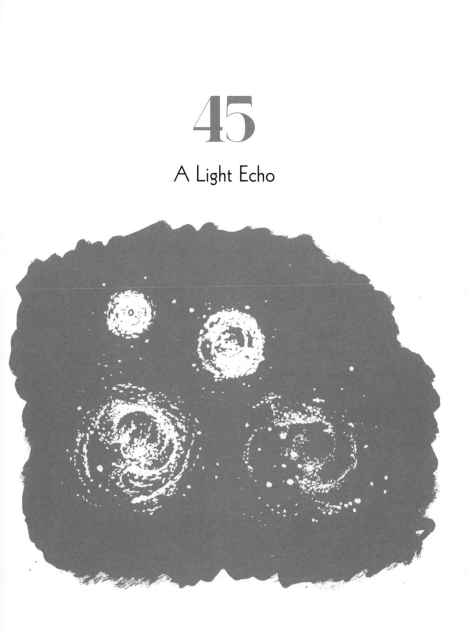

Once upon a time, deep in a rather unremarkable constellation, was a rather unremarkable star. But in January 2002 something rather astonishing happened: the star became extremely bright— about ten thousand times brighter than it had been just a few weeks before. The star is V838 Monocerotis, the 838th variable star in the constellation Monocerotis (the Unicorn). Perhaps even more impressive than this huge change in brightness is what the light allowed us to see. Shortly after the star increased in luminosity, the Hubble Space Telescope took a look. Hubble's tremendous powers of resolution revealed a messy halo of dust surrounding the star, and when the telescope snapped additional images in the years that followed, this halo appeared to grow. In fact, it wasn't the halo of dust itself that was growing. Instead, the phenomenon is called a light echo. As the flash of light from the star traveled outward, it illuminated surrounding dust, allowing this material to be seen all the way from Earth.

Astronomers are still trying to work out exactly what happened to make the star bloom in brightness so spectacularly. One theory is that V838 Monocerotis was a supergiant star in its death throes; another is that two stars merged into one with a spectacular flash; while another speculates that the star had one or more giant planets, and that the flash came from the star "eating" the planet(s)!

To see the light echo you'll need two things: the Hubble Space Telescope, and a time machine.

1. The Hubble Space Telescope trained its sights on the star in May 2002 and saw a clear ring of gas and dust around the star.

2. In September 2002 it took another look, and more of that gas and dust was now visible. The light echo had grown.

3. Another Hubble image from February 2004 shows an even bigger and more detailed light echo.

4. When Hubble looked again in November 2005, the light echo appeared more ghostly and dispersed.

5. If you were to look today, the light echo would be so dispersed and so faint that you wouldn't see it. It's now beyond the powers of even the Hubble Space Telescope to see.

46

A White Flag

Have you ever seen the iconic photographs of the Apollo astronauts on the surface of the Moon? The ones where they are standing by their rovers, or surrounded by their own footprints in the lunar dust? The chances are that there was an American flag clearly in shot. Each of the six Apollo missions that landed humans on the Moon planted a flag on the lunar surface, and the astronauts weren't shy about using the flags for photo opportunities. But while the images of these history-making missions are as vivid as ever, the flags themselves have lost their luster. Thanks to the Moon's lack of atmosphere, the red and blue colors of the flag have been bleached to white, the result of ultraviolet radiation and extreme temperature variations. The lack of atmosphere means the moon has no protection from micro-meteor impacts, so the flags probably have a few holes in them now too.

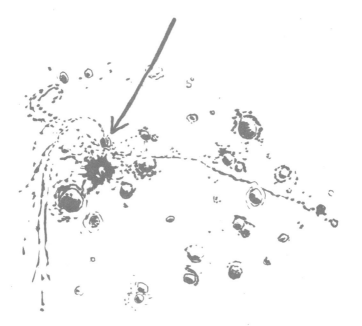

1. Have a look at images taken by the Lunar Reconnaissance Orbiter (LRO). This spacecraft launched in 2009 and has allowed scientists to map the surface of the Moon in unprecedented detail.

2. If you focus on LRO images of the Apollo 12, 16, and 17 landing sites, and compare photographs taken at different times (i.e. with different Sun angles), you can see the shadows of the flags and flagpoles! This proves that at least three of the flags are still flying—although their colors are likely to be long gone.

3. The best way to see what condition the flags are in now is to get yourself all the way to the Moon and take a look!

• TOP FACT! •

Astronaut Buzz Aldrin remembers blasting off from the Moon toward home and seeing the flag his Apollo 11 crew planted fall over thanks to the rocket blast. However, some of the other Apollo flags are believed to be still standing today!

47

Galaxy Groups and Clusters

As big as galaxies are, they are not the biggest structures in the Universe. Because what's bigger than a galaxy? A group of galaxies. The Milky Way is part of a group of galaxies imaginatively named the Local Group. This galactic super-squad includes the Andromeda Galaxy (see page 80), the Large and Small Magellanic Clouds (see page 76), as well as the Triangulum Galaxy, and about thirty more on top of that. Most of the galaxies in the Local Group are dwarf galaxies, and some of them are so small and faint that there may be others in the group that have not yet been discovered. If a group of galaxies is big enough, it's called a cluster. These are huge numbers of galaxies bound together by mutual gravitational attraction. Our Local Group of galaxies is part of a galaxy cluster called the Virgo Supercluster, which stretches over a bewildering one hundred million light-years of space.

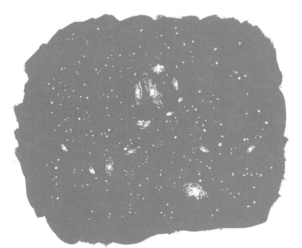

1. Find the constellation of Virgo. One way to do this is to start at the Plough/Big Dipper and follow the mnemonic "arc to Arcturus, spike to Spica." This means draw an arc through the 'handle' of the Big Dipper and keep going until you reach the bright star Arcturus, then continue that arc further to reach the brightest star in the Virgo constellation, Spica.

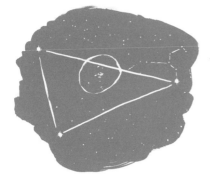

2. Use Spica and Arcturus to form a triangle with a third star — Regulus, in the constellation of Leo. Roughly in the middle of this triangle, you'll find the Virgo Cluster. This forms the heart of the Virgo Supercluster.

3. Sweep your telescope around this area and you should bump into several galaxies. Once you've found your way to this area, you can return again and again to find different component galaxies within this enormous group. It's pretty amazing that you can see some of these, given that they are around sixty-five million light-years distant!

Quasars

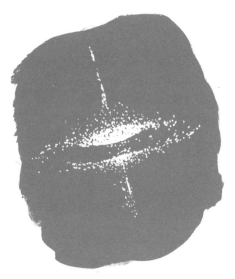

Quasars (short for Quasi-Stellar Objects) are among the most distant known objects in the Universe. They are "quasi stellar" because they resemble stars when viewed from Earth, but they are very different and much more exciting! Quasars are intensely powerful centers of wildly distant galaxies. They are incredibly far away—as much as thirteen billion light-years away from Earth. When we look at a quasar, we are glimpsing the deep past—a time closer to the Big Bang than the time we live in today. Quasars shine hundreds or even thousands of times brighter than an entire galaxy like the Milky Way! This makes them some of the most luminous objects in all the Universe. Astronomers believe that at the center of each quasar is a supermassive black hole that helps supply its energy.

1. Quasars are *so* far away that despite their incredible luminosity, they appear disappointingly faint; far too faint to be seen with the naked eye.

2. It *is* possible to see quasars with backyard telescopes, but only the very best ones, and only if the conditions are just right. You would need a telescope with an aperture of at least eight inches (200 mm)—which is a fair bit bigger than the average hobby telescope—but even then, quasars will just look like faint stars.

3. A better bet is to gain access to a world-class telescope, or just look at pictures it has already taken. Look for images of bright halos surrounding distant quasars, or brilliant jets of X-rays darting thousands of light-years out from the center of a quasar.

49

Pillars of Creation

It's a pretty grand title, but it's fitting. Located in the Eagle Nebula, these towering tendrils of gas and dust are stellar nurseries, where new stars and solar systems form. These colossal clouds are about five light-years from top to bottom, or roughly thirty trillion miles. To put that in perspective, our entire Solar System would fit in one of the tiny, finger-like protrusions, poking up from the top of the pillars. The pillars are illuminated by bright stars above, but strong winds of ionized gas produced by these young stars are gradually eroding the pillars, so be glad you are here now to see them in all their glory.

The Pillars of Creation are so far away that only the largest telescopes under the best viewing conditions are able to see them. However, the nebula in which they reside—the Eagle Nebula—can be seen with a small telescope, and is best viewed in July and August.

1. In the northern hemisphere in summertime, look for the constellation Sagittarius, and the asterism known as the Teapot (see page 99).

2. Draw a line from the brightest star in Sagittarius—Kaus Australis (also called Epsilon Sagittarii)—to just east of the star Kaus Media (also called Delta Sagittarii) and keep going for about four times that distance.

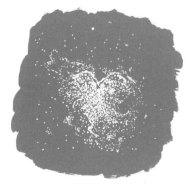

3. Using a low-powered telescope or binoculars, you should be able to see a cluster of small stars, grouped together—this is the Eagle Nebula. But watch out! This area of the night sky is rich in nebulae and deep sky objects, including the Lagoon Nebula, Trifid Nebula, and Swan Nebula (see page 98).

50

Hubble Deep Field

The Hubble Deep Field images are some of the most mind-blowing pictures ever captured. Taken by the Hubble Space Telescope, these images reveal the dimmest and most distant objects in space. The telescope was pointed at a tiny region (only 1/24 millionth of the whole sky) that appeared devoid of stars to the naked eye, and it started taking pictures. When the pictures were combined, an awesome image appeared. Almost every tiny speck of detail in these beautiful scenes shows an entire galaxy. The first Hubble Deep Field image revealed close to three thousand galaxies; some of them bright, some dim, some red, some blue, some big, some small—but each with the potential to be home to thousands of millions of stars.

The light from these galaxies is so dim because they are incredibly far away. The Hubble Deep Field images reveal some of the most distant objects in the night sky. The light from these faraway spectacles takes so long to get to us, that by gazing at these images we are peering back in time—the galaxies we see are from an earlier stage in the development of our Universe. By looking at these stunning images, we are time-traveling.

1. Borrow the powerful Hubble Space Telescope (or visit the Hubble Space Telescope website from the comfort of your own home—see page 143).

2. Deliberately point it at a black, seemingly empty patch of space (the first deep field aimed close to the constellation of Ursa Major).

3. Take up to two thousand long-exposure photographs of this "nothingness" and combine them into one image.

4. Look at the resulting picture and be flabbergasted—even the most seemingly sparse areas of space are teeming with beautiful, distant structures. We live in a big, busy, beautiful Universe.

Glossary

Here is a list of terms that are commonly used in astronomy. Many more are explained throughout the book.

Asterism A distinct pattern of stars that is not a constellation; this can lie within a constellation (e.g., Ursa Major) or between constellations (e.g., the Summer Triangle).

Binary system Two stars that revolve around a common center of mass and are linked by the same center of gravity.

Celestial pole Imaginary points in the sky in the northern and southern hemispheres where the axis of the Earth's rotation, if extended, would touch the celestial sphere (see below).

Celestial sphere An imaginary sphere around the Earth within which the objects seen in the night sky appear to lie.

Circumpolar star A star that always stays above the horizon when viewed from a particular location.

Star cluster A group of stars bound together by gravity.

Comet A body of ice and dust that travels around the Sun.

Constellation One of eighty-eight named patterns of stars into which the night sky is divided.

Galaxy A giant system comprising stars, gas, and dust held together by gravity.

Globular cluster A large group of old stars with a spherical form.

Latitude A notional circle drawn parallel to the Earth's equator, measured in degrees north or south.

Light-year A unit equivalent to the distance that light travels in one year.

Magellanic Clouds Two irregular dwarf galaxies, the Large and Small Magellanic Clouds, visible near the south celestial pole.

Magnetic North Pole The point on Earth that a compass needle points to. This is near the geographic North Pole but the exact location varies over time.

Maria Smooth, dark-colored areas on the surface of the Earth's Moon.

Meteor Space debris burning up as it enters the Earth's atmosphere. Also known as a shooting star.

Meteor shower This occurs when a number of meteors enter the Earth's atmosphere from the same direction and at about the same time.

Meteorite A piece of space debris that reaches the surface of the Earth intact.

Nebula A giant cloud of gas and dust in space.

Neutron star The compressed core of a collapsed giant star that consists almost entirely of neutrons.

Opposition When a planet appears exactly opposite the Sun in the sky as seen from Earth.

Photosphere The visible surface of the Sun.

Planetary system A star with planets (and perhaps moons, asteroids, and comets, etc., as well) in orbit around it.

Pulsar A spinning neutron star that emits bursts of radio energy.

Quasar A bright object deep in space that emits huge amounts of energy.

Refraction The bending of light as it enters a new medium (e.g., when light travels from space to Earth's atmosphere).

Shooting star See meteor.

Supergiant star A star that has expanded to a huge size and is likely to explode as a supernova.

Supernova The explosion of a star. This may be due to a star accumulating too much matter from a companion star within a binary system (see previous page) or when a star has reached the end of its life cycle and has run out of fuel.

Terminator The line between the dark and light sides of a planet or moon.

Totality The period of time during a solar eclipse when the Sun is completely obscured.

Zenith The point directly above the observer.

Sources

Here are some useful apps and websites that can help you map the night sky and spot particular stars and planets, and which provide all the latest news and information.

Phone apps

Stellarium

The Stellarium Mobile Sky Map is like a little planetarium in your pocket. It has a realistic and accurate night sky map, and is one of the best apps for finding planets and constellations in the night sky.

Star Chart

This is a free app that gives you your own personal star chart in the palm of your hand. It allows you to turn your phone into a window to the night sky.

Websites

www.skyandtelescope.com
This monthly American magazine has a fantastic website with top tips for stargazers.

www.skyatnightmagazine.com
The BBC Sky at Night magazine also has an amazing website packed with advice for beginners and experts alike. A great place to look before you buy a telescope!

www.nasa.gov
The pioneers of space exploration and scientific discovery! For the latest news, images, and video from space—look no further.

www.earthsky.org
A fantastic resource for stargazers, this website is updated daily and boasts tons of astronomy essentials.

www.heavens-above.com
Head to this website for their printable interactive star chart function—a useful alternative to apps, if paper is more your thing.

www.spotthestation.nasa.gov
Want to find out when the International Space Station can be spotted from your location? This is the website to check.

www.hubblesite.org
The official website for the Hubble Space Telescope, with some of the most breathtakingly beautiful images of our universe you'll ever see.

Acknowledgments

Firstly, I'd like to thank the fine folk at Pavilion: Michelle Mac, Bella Cockrell, and particularly the inimitable Krissy Mallett for her unwavering positivity through the whole process, and for making it all happen in the first place! Huge thanks and wide-open eyes of amazement go to Maria Nilsson for her charming (and ludicrously speedily-done!) illustrations. Supermassive thanks go to Dr. Darren White for his invaluable input—cheers duck! And a big thank you goes to the Physics and Astronomy department of The University of Sheffield, for deepening my love of the skies all those years ago. I also want to thank all of my friends and family for their support and understanding; particularly my wonderful mother for her early proofreading (that's three decades of checking errant apostrophes now, mom!), my brother David, who is a far better writer than I will ever be, and my Dad, for helping me climb onto the shed to watch the Hale–Bopp comet when I was a little kid in Birmingham. One night of staring into the sky can stay with you a lifetime. Finally, and most importantly, my loudest thanks go to my biggest supporter, the largest planet in my solar system—Jove. I love you.